Max Leonard
Camille McMillan

BUNKER RESEARCH

îsola

« Over the years I had puzzled out a good deal in my mind, but in spite of that, far from becoming clearer, things now appeared to me more incomprehensible than ever. The more images I gathered from the past, I said, the more unlikely it seemed to me that the past had actually happened in this way or that way, for nothing about it could be called normal: most of it was absurd, and if not absurd then appalling. »
Vertigo
WG Sebald, 1990

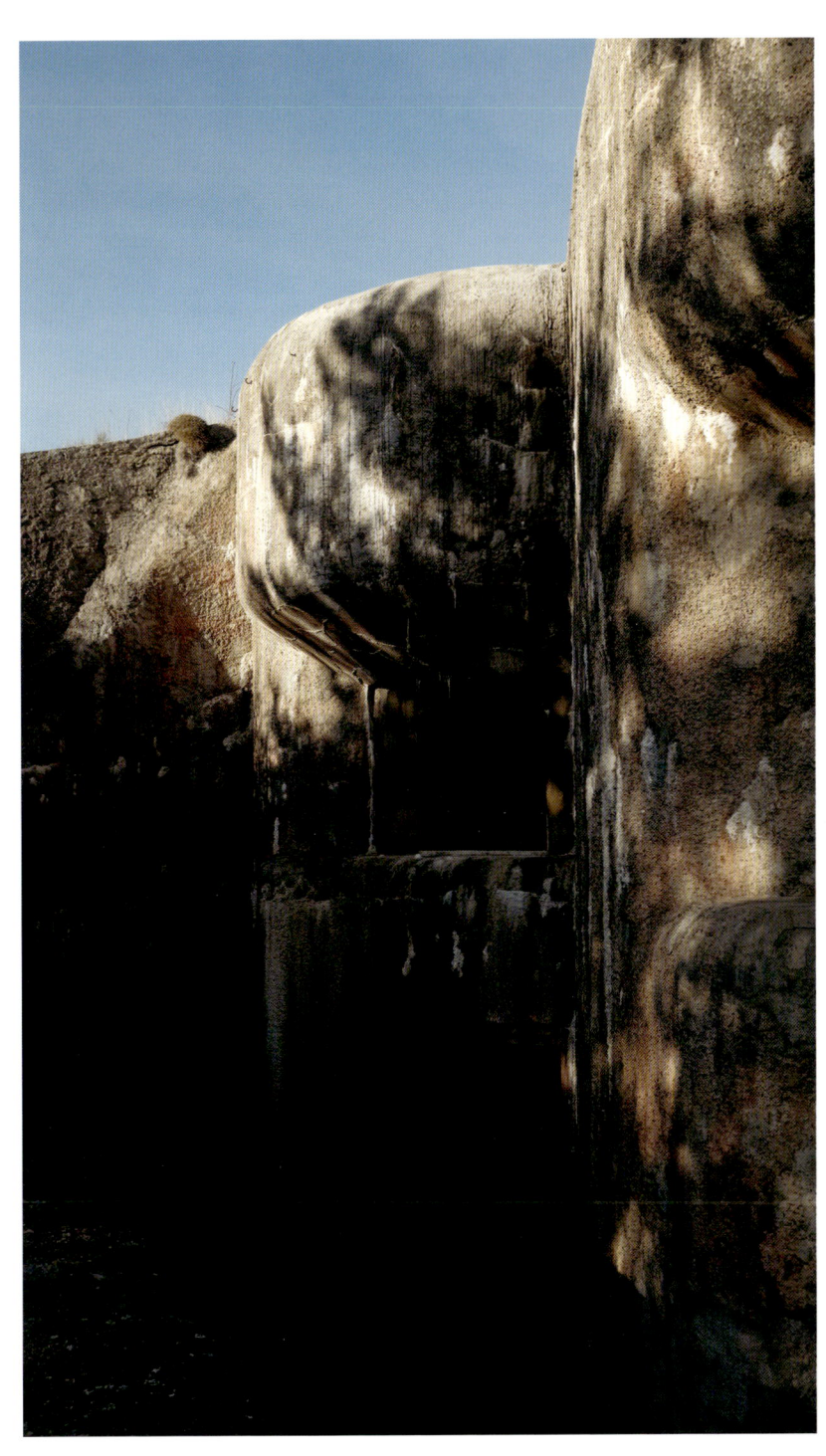

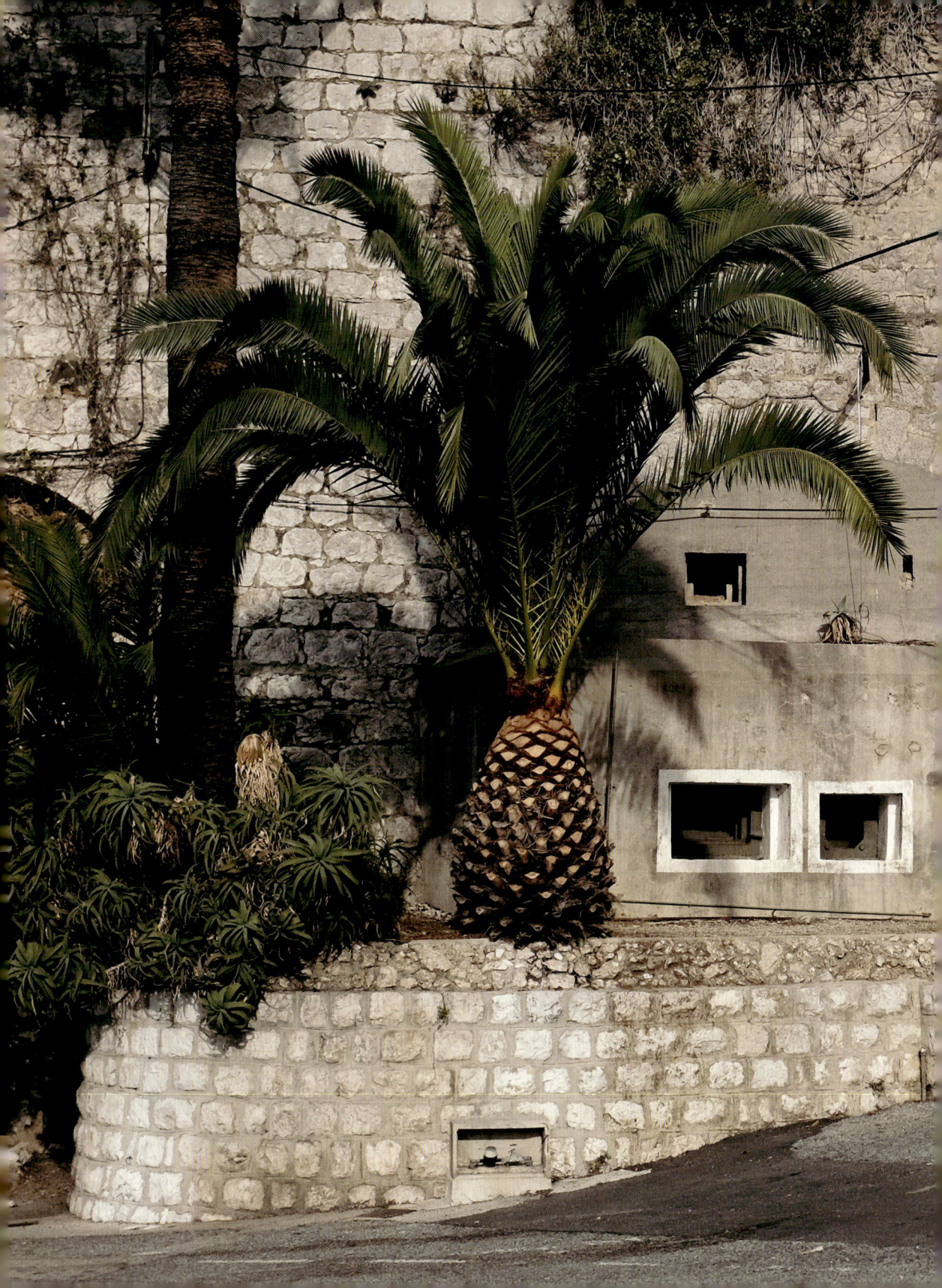

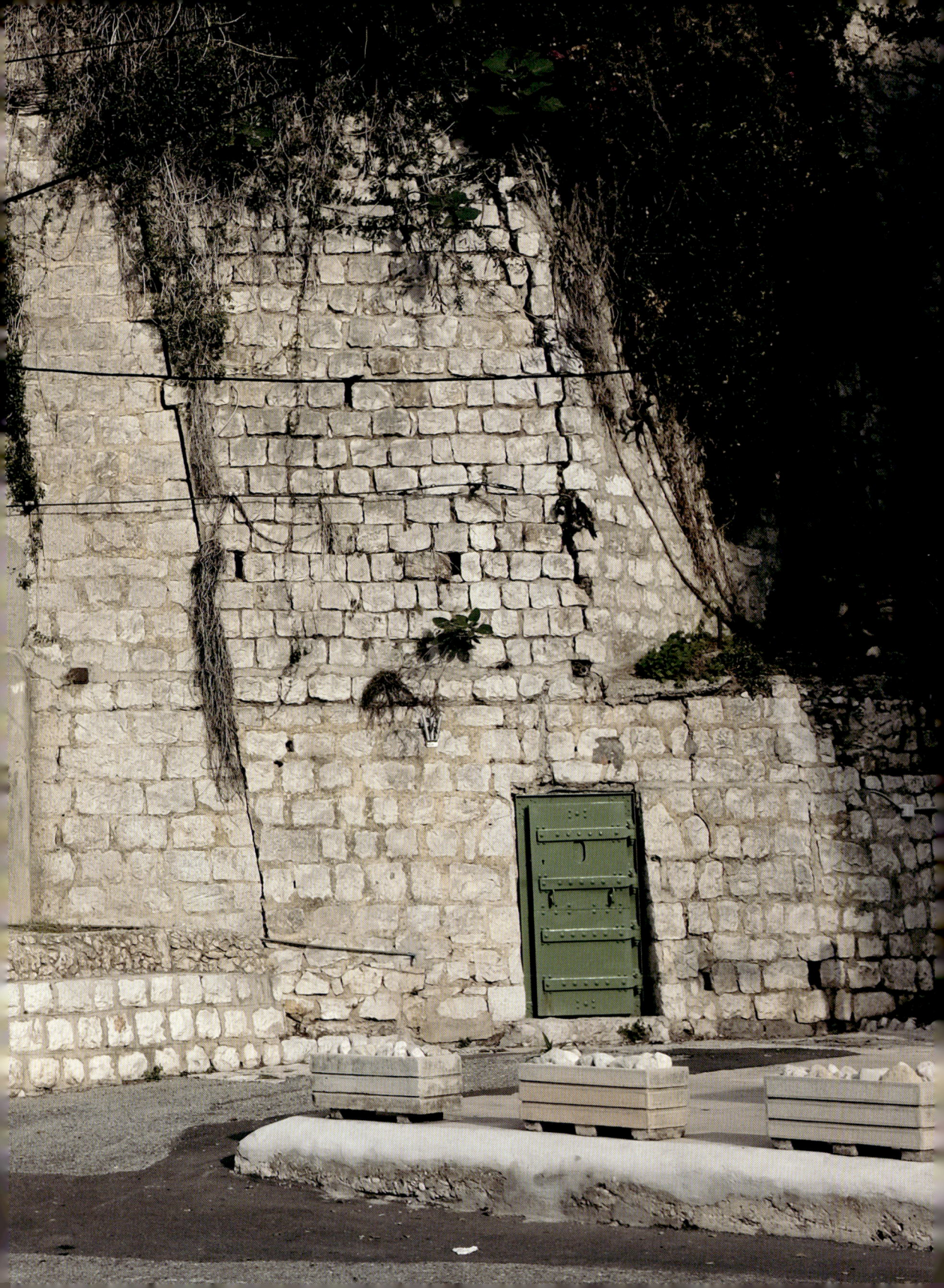

Previous pages
Avant-poste du Pont Saint Louis
SFAM / Corniches
One officer, seven men

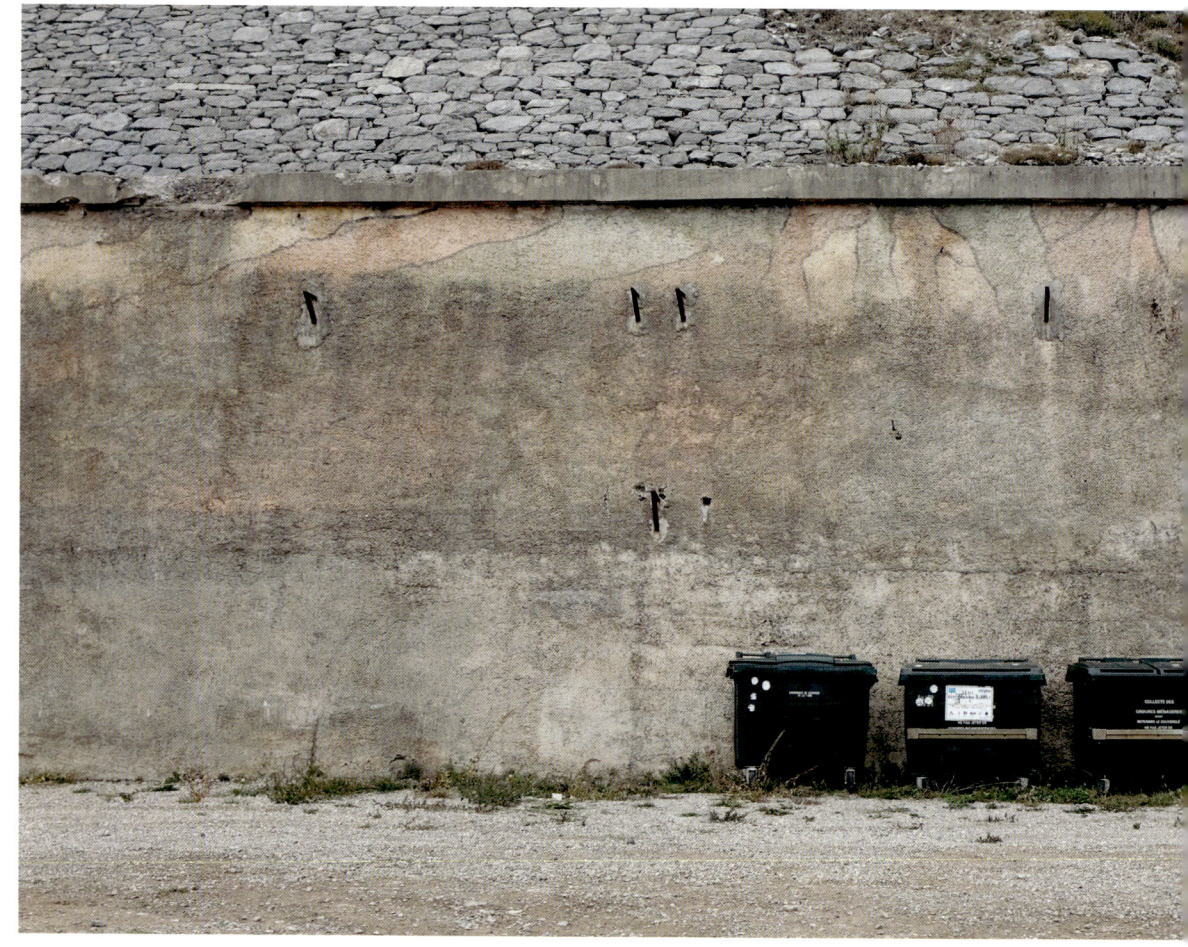

Below
Gros ouvrage de Rimplas
SFAM / Tinée-Vésubie
Eight officers, 334 men

Right
Petit ouvrage de Frassinéa
SFAM / Tinée-Vésubie
Three officers, 33 men

Following pages
Gros ouvrage de Roquebrune
SFAM / Corniches
Nine officers, 284 men

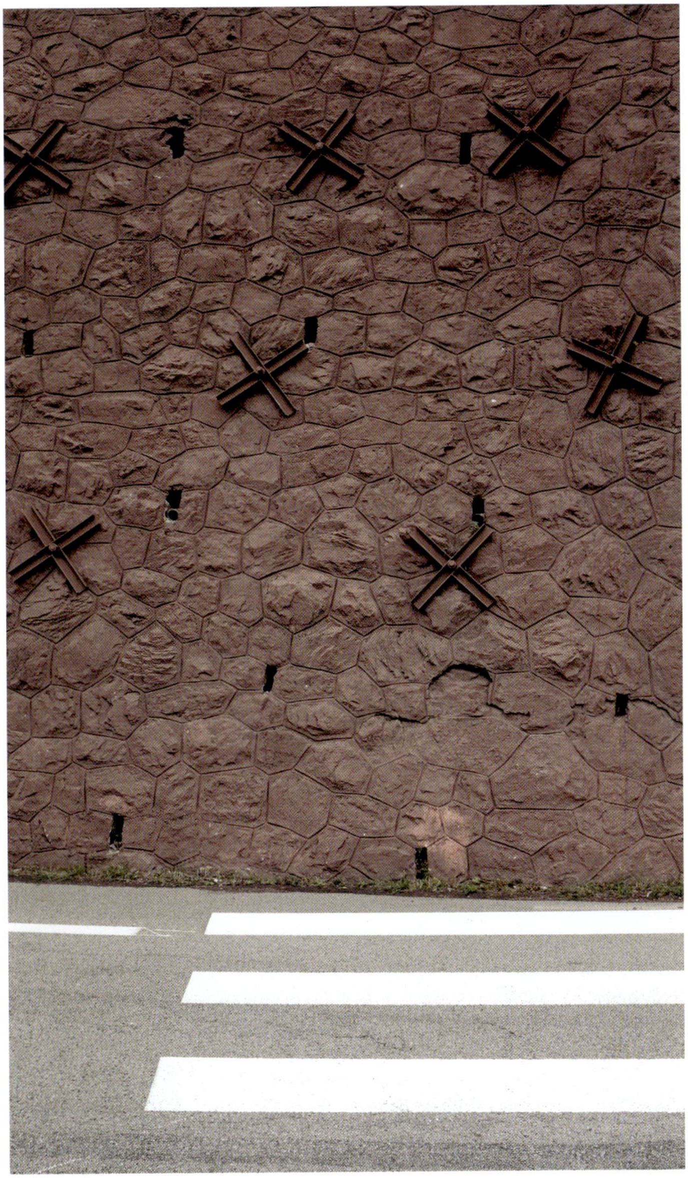

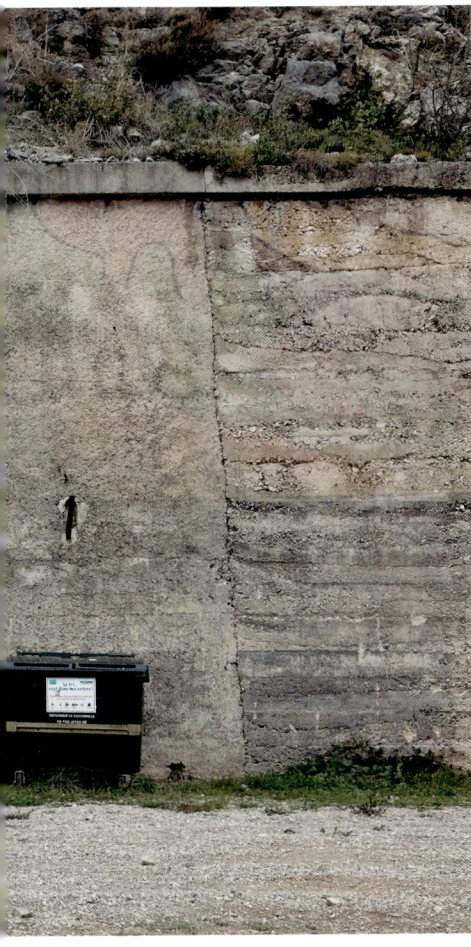

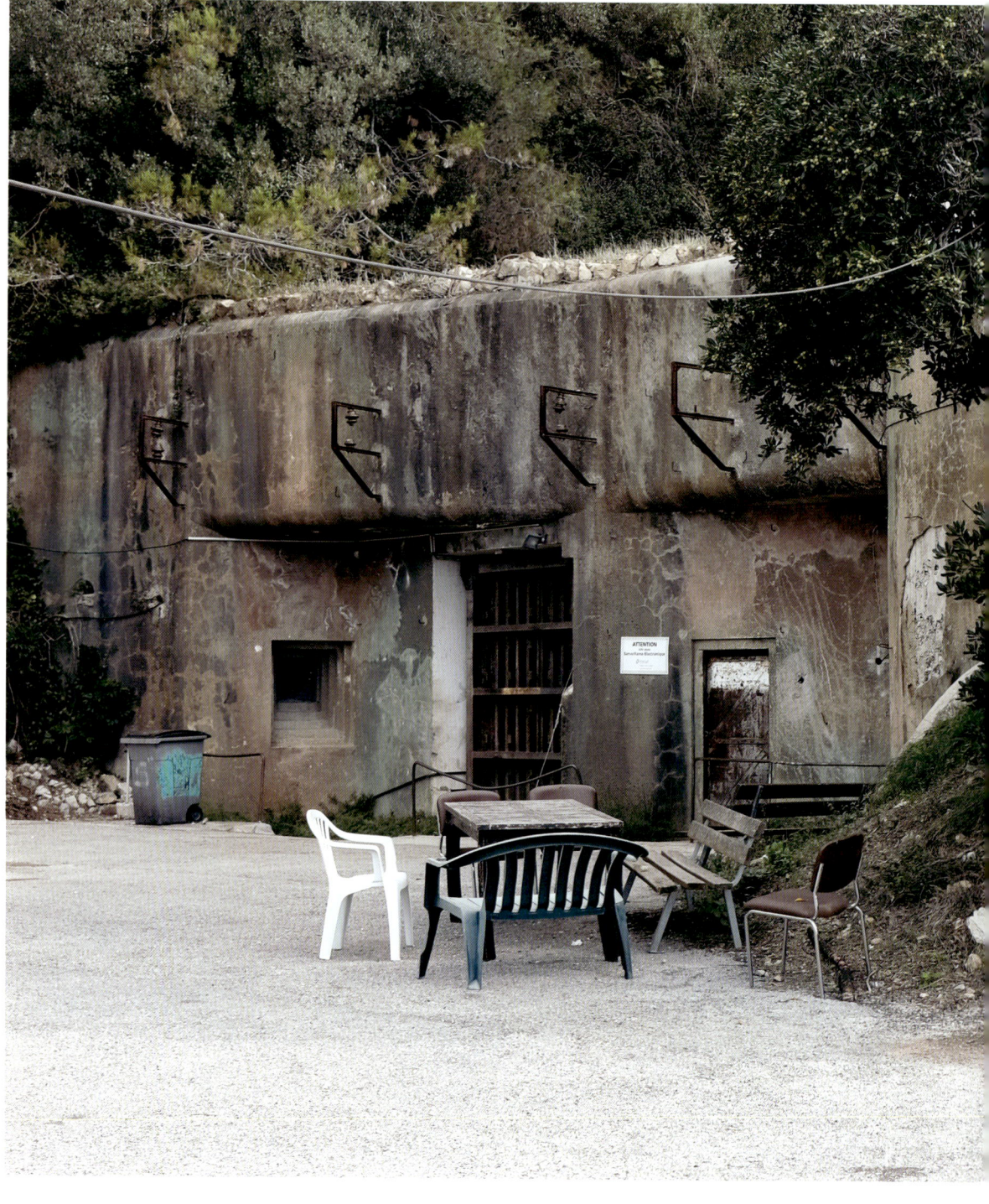

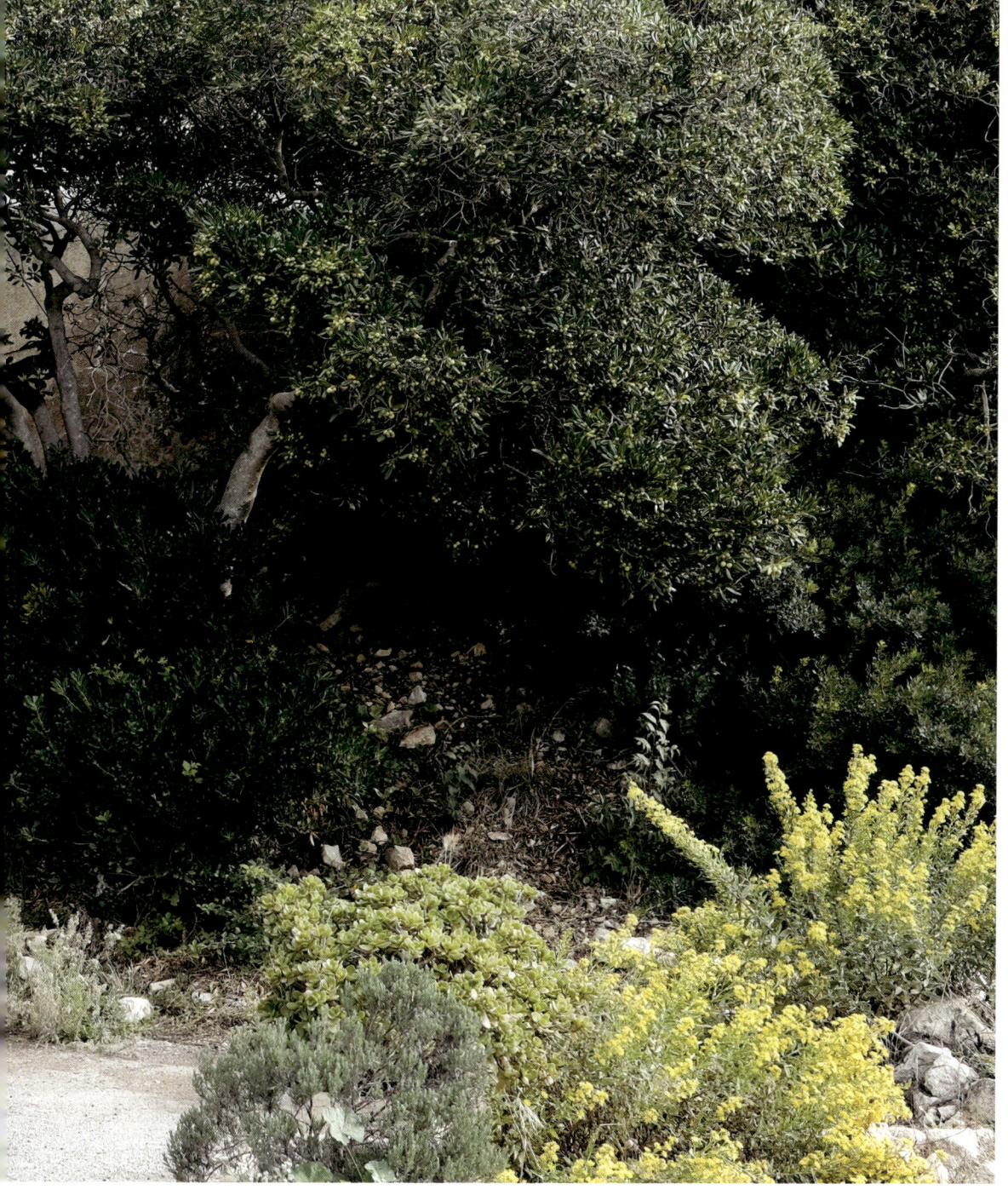

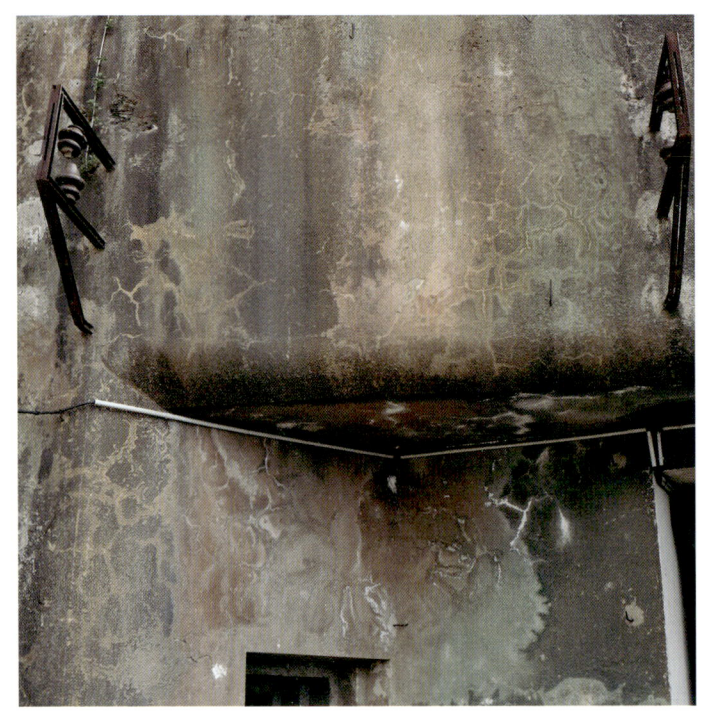
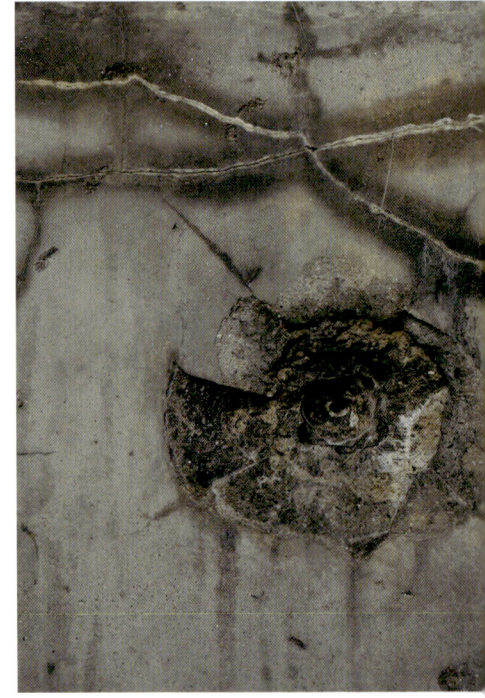

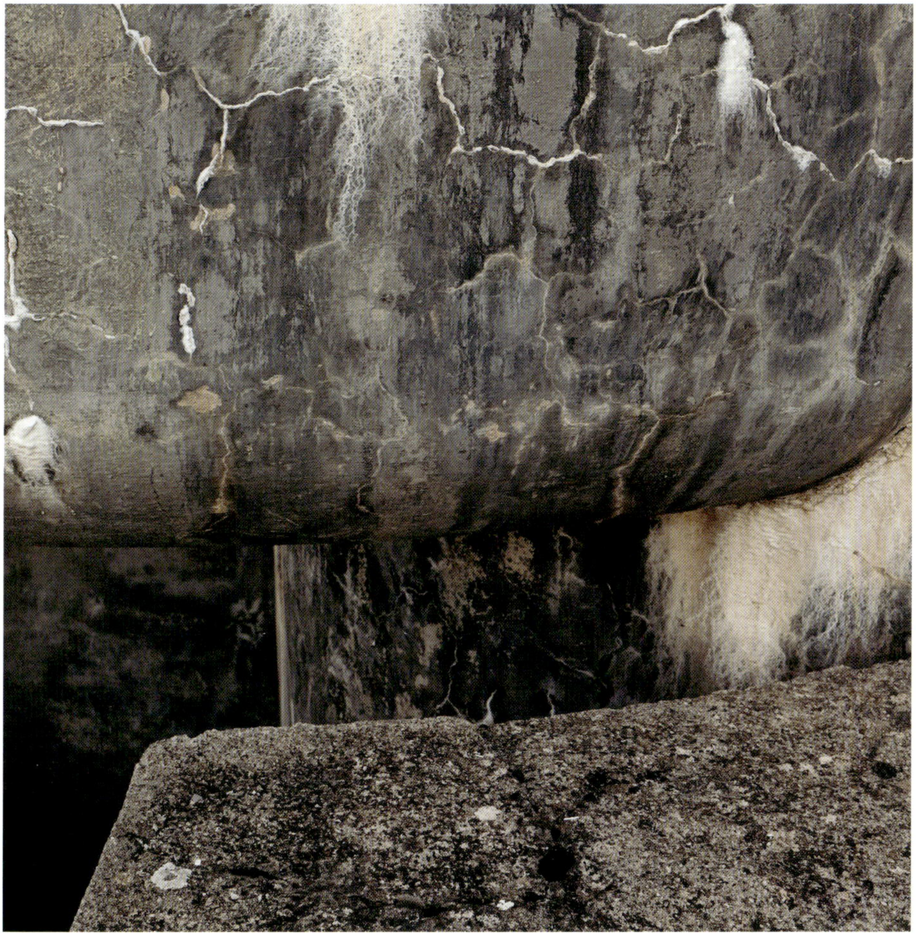

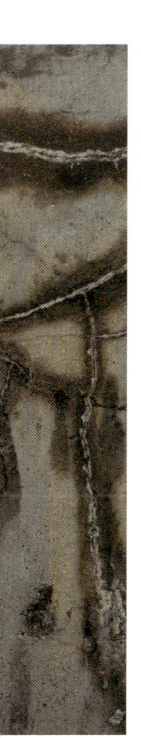

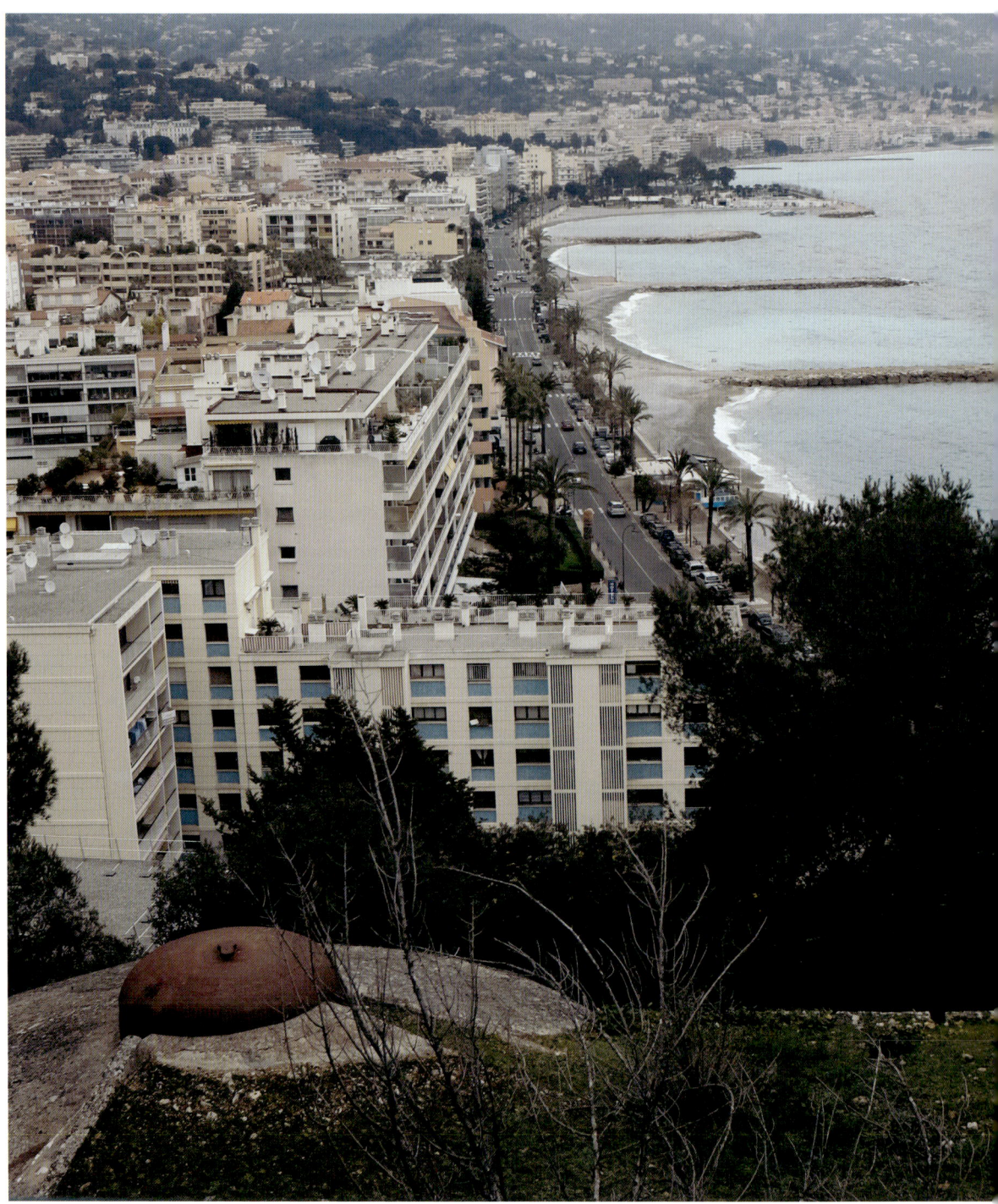

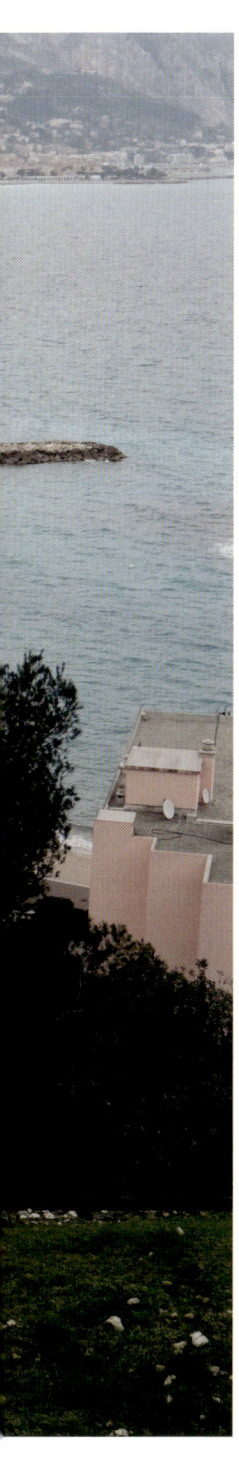

Below
Casemate Abéliéra
SFAM / Tinée-Vésubie

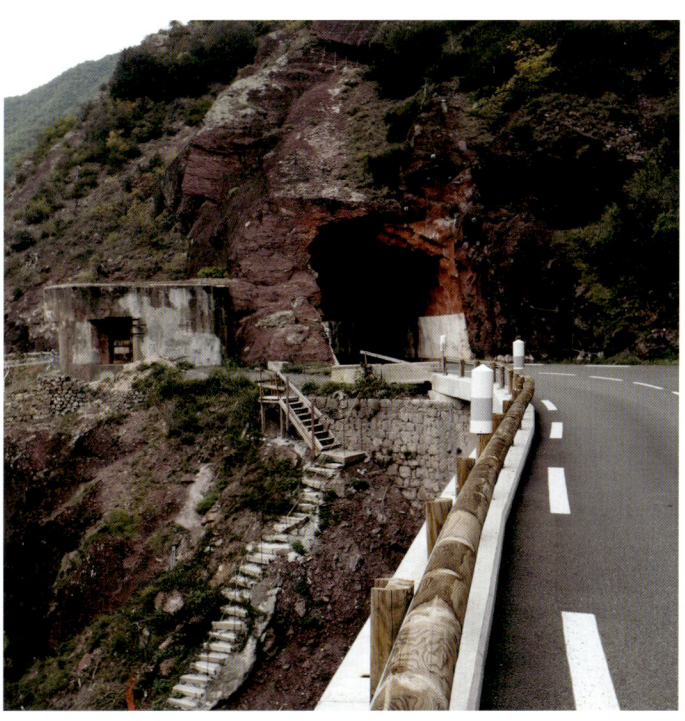

Left
Gros ouvrage du Cap Martin
SFAM / Corniches
Seven officers, 353 men

Positioned at the extreme south of the Maginot Line, the *ouvrage* overlooked the town of Menton and the border only two kilometres away. When the Italians attacked on 20 June 1940, they took Menton but did not breach the Maginot defences. The bunker was blown up from the inside in 1944, by German soldiers as they retreated.

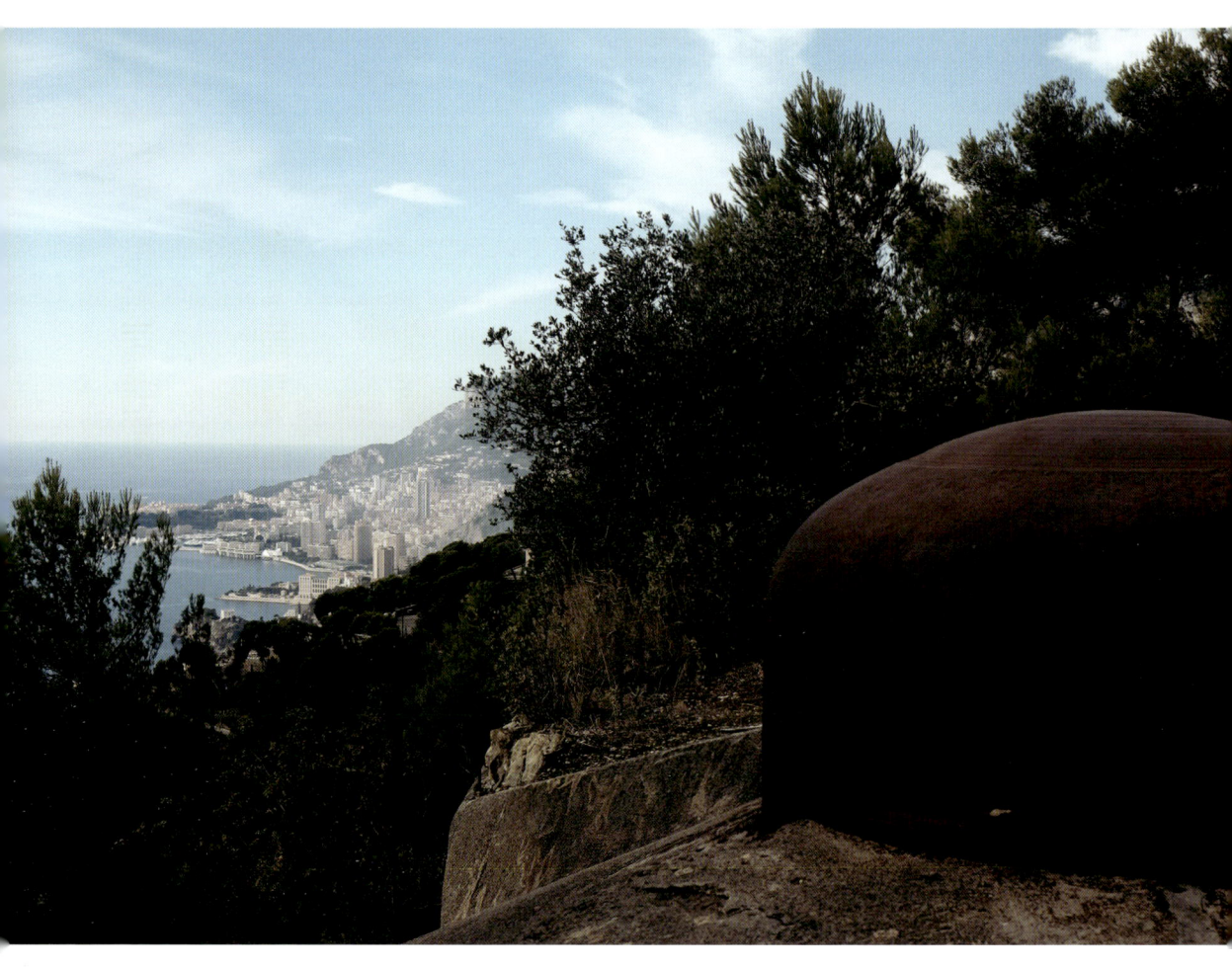

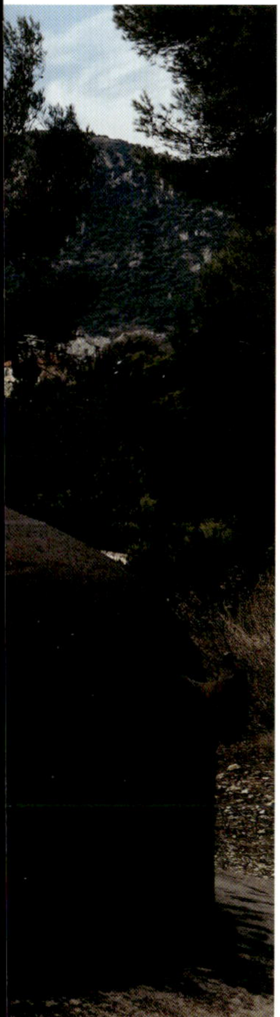

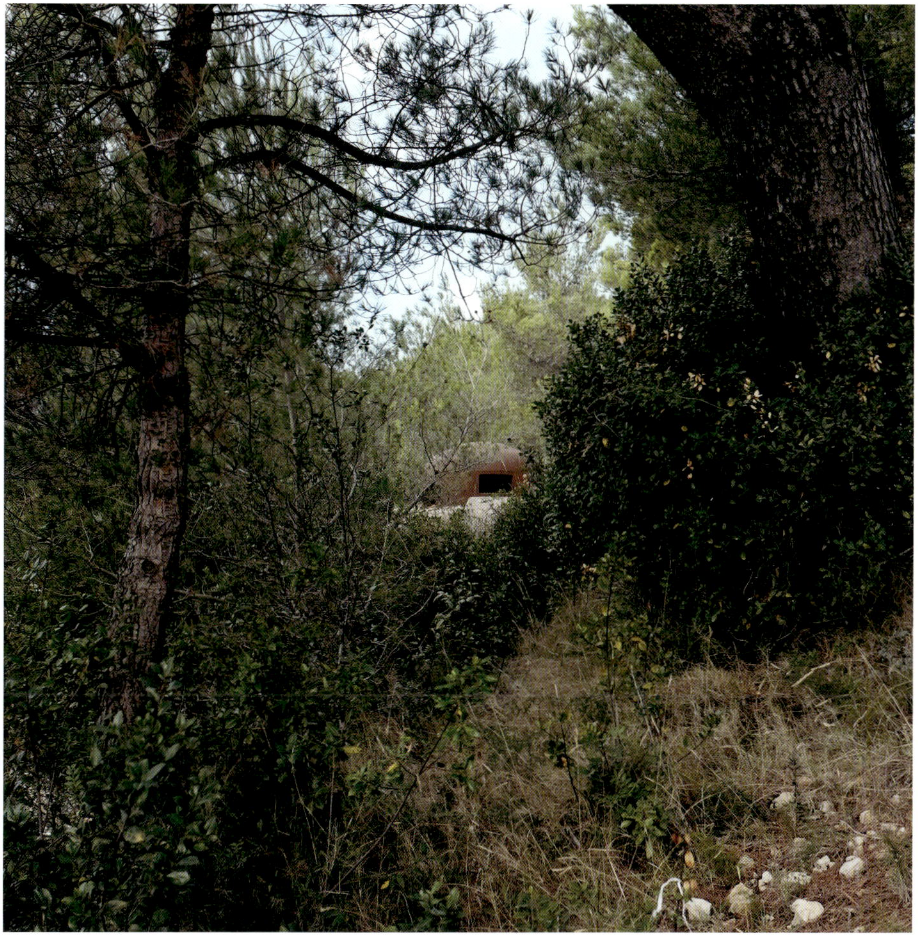

Gros ouvrage de l'Agaisen
SFAM / Sospel
Seven officers, 295 men

Mont Agaisen dominates the Sospel basin.
This bunker complex is now open for
guided tours and is used as a launch
site for paragliders.

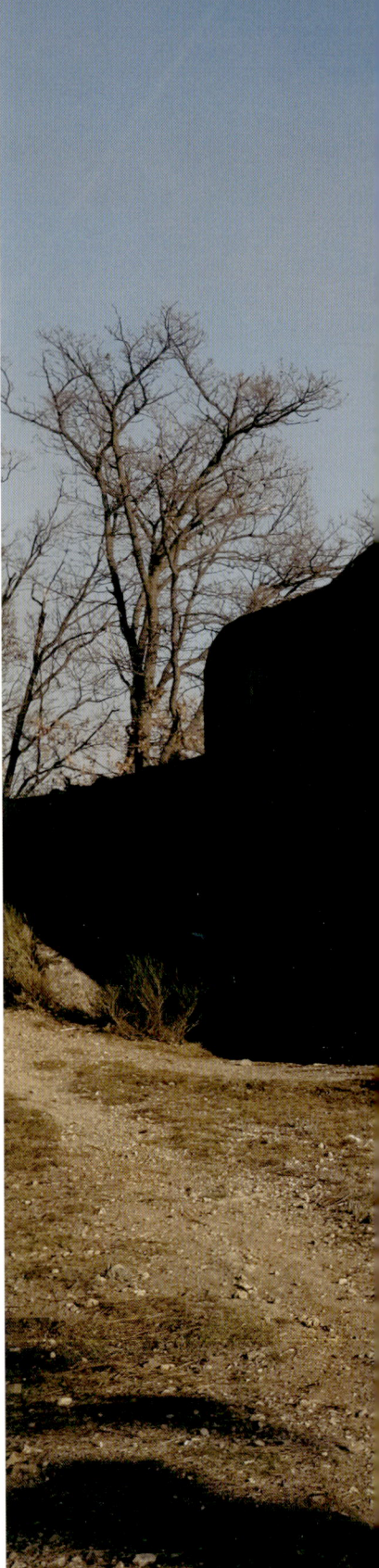

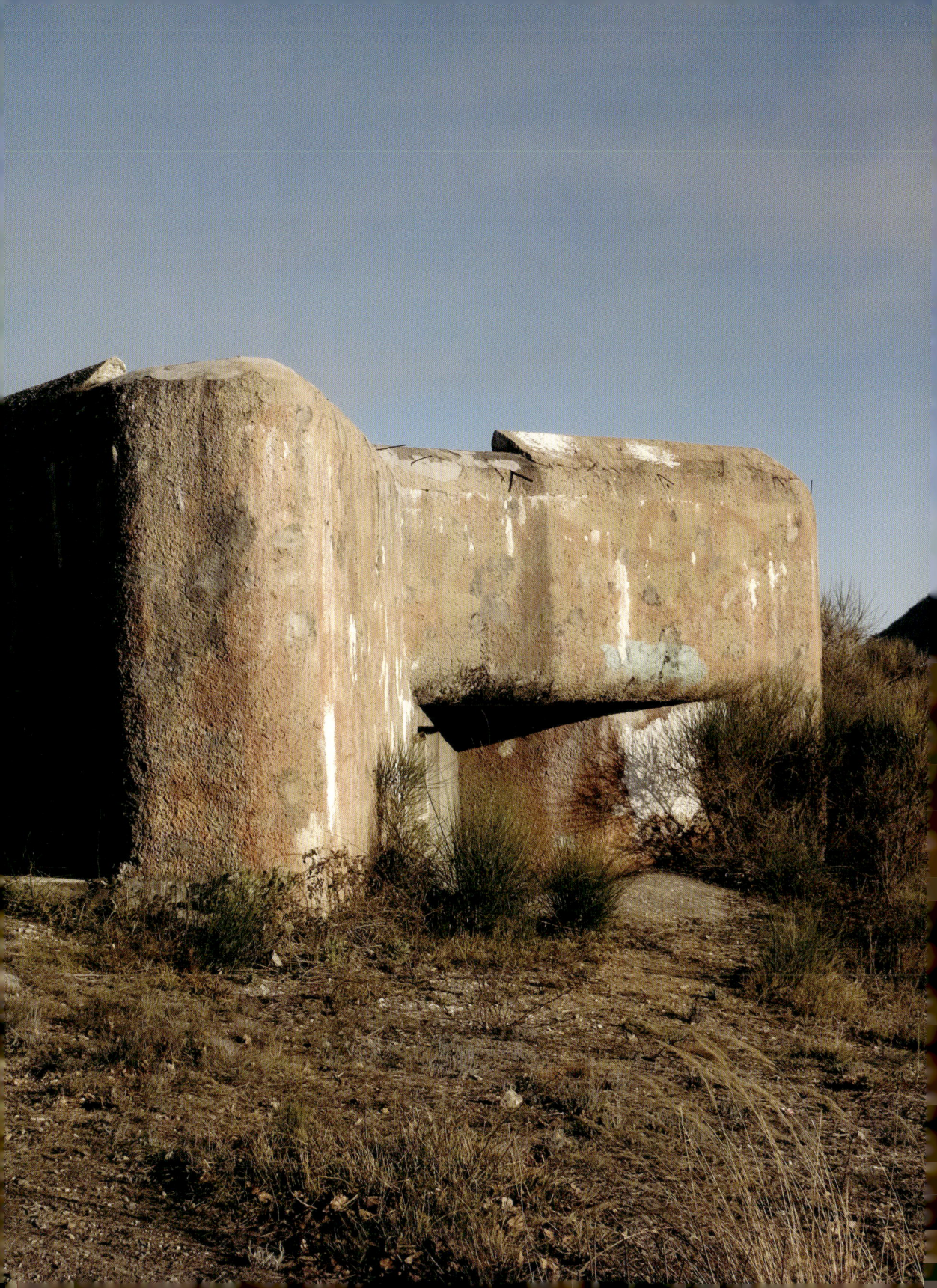

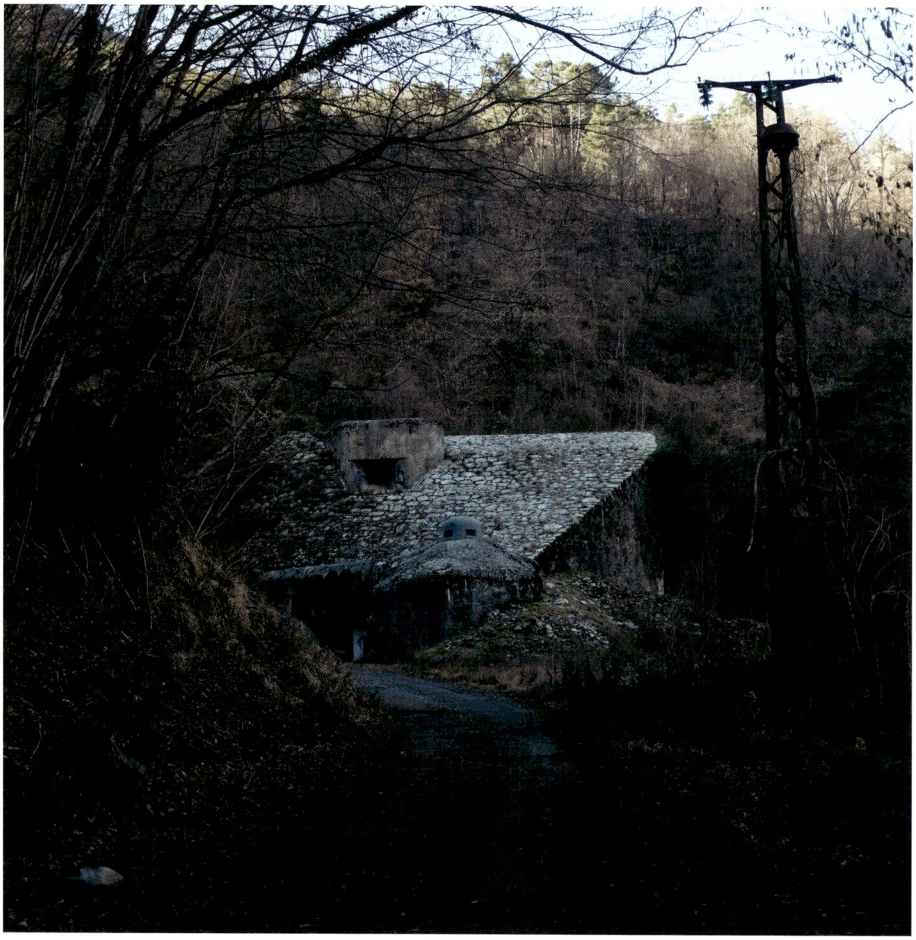

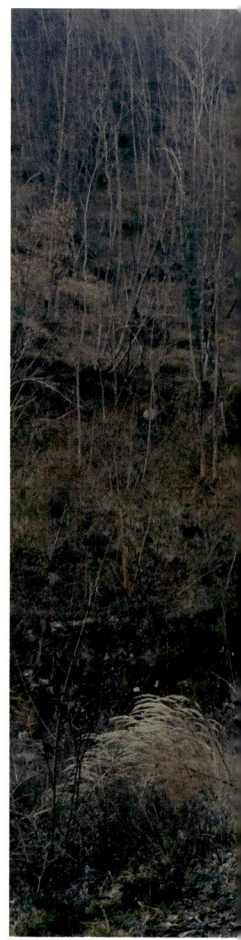

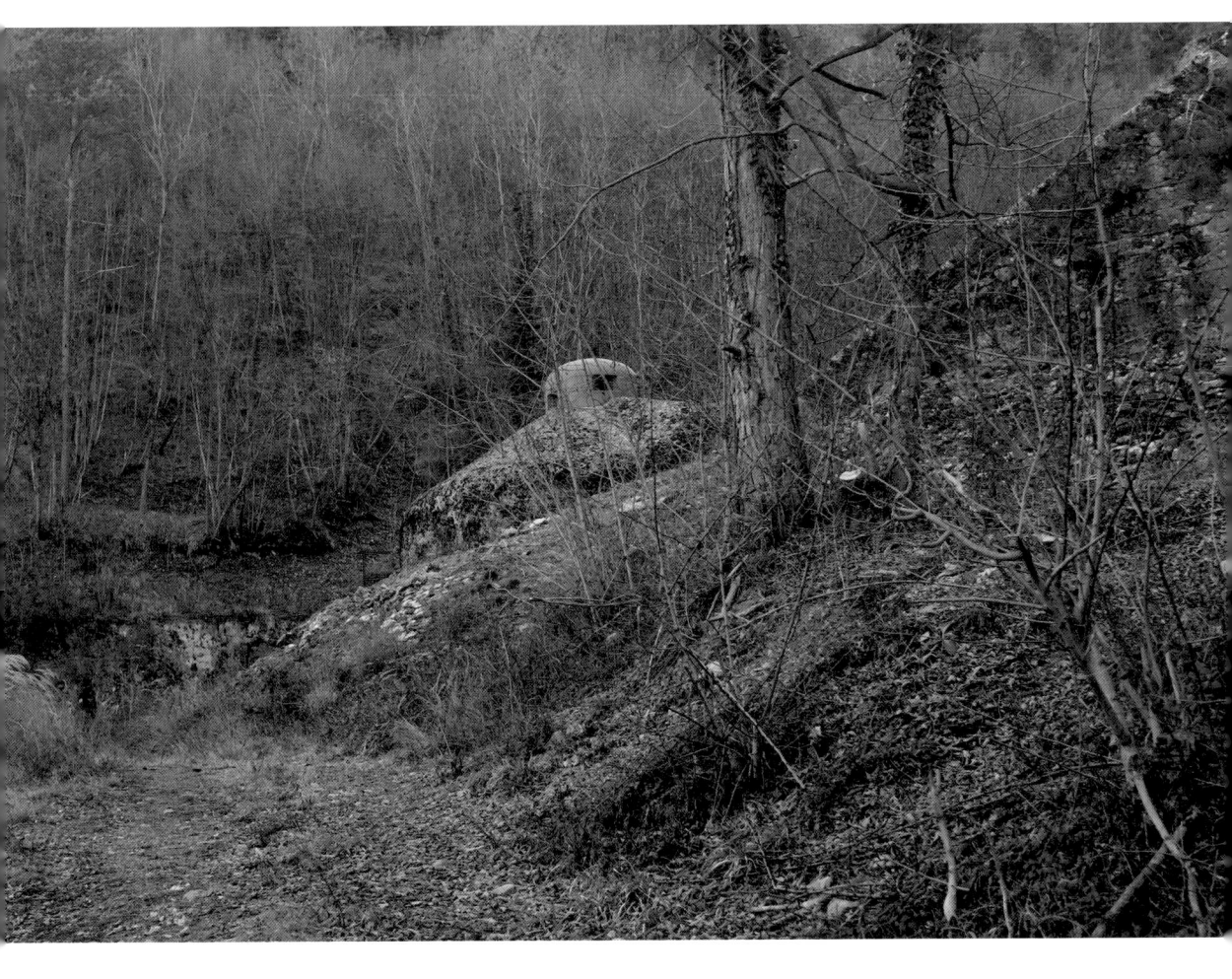

Gros ouvrage de Gordolon
SFAM / Tinée-Vésubie
Five officers, 246 men

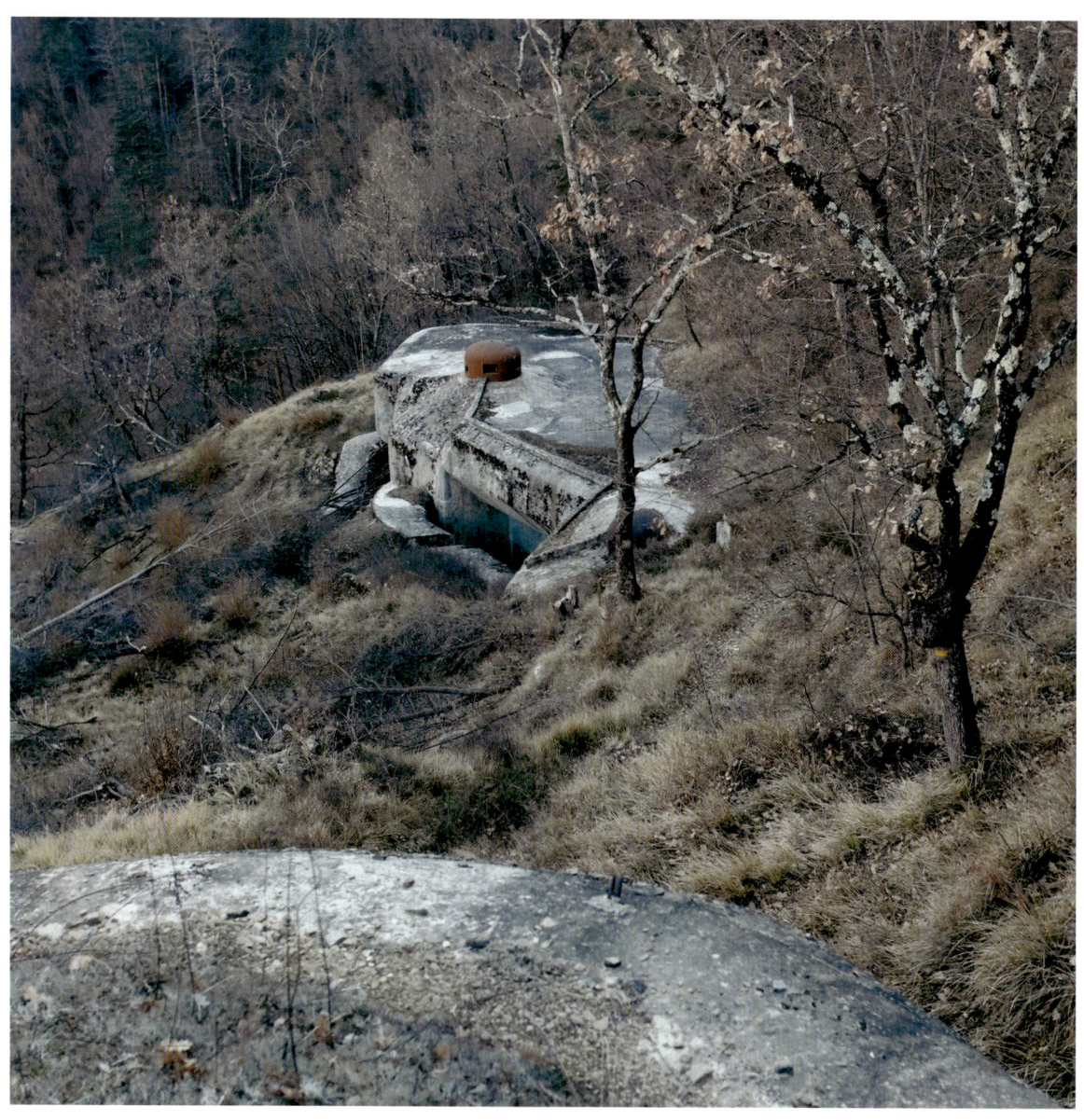

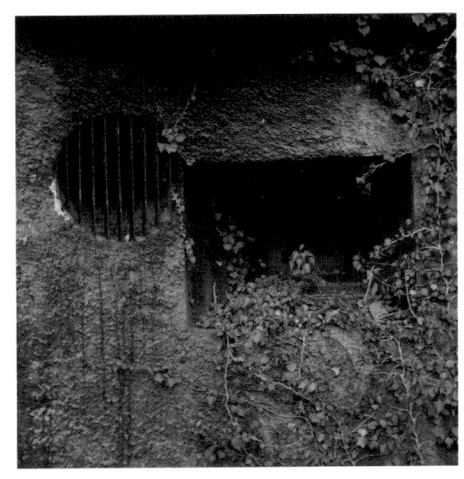

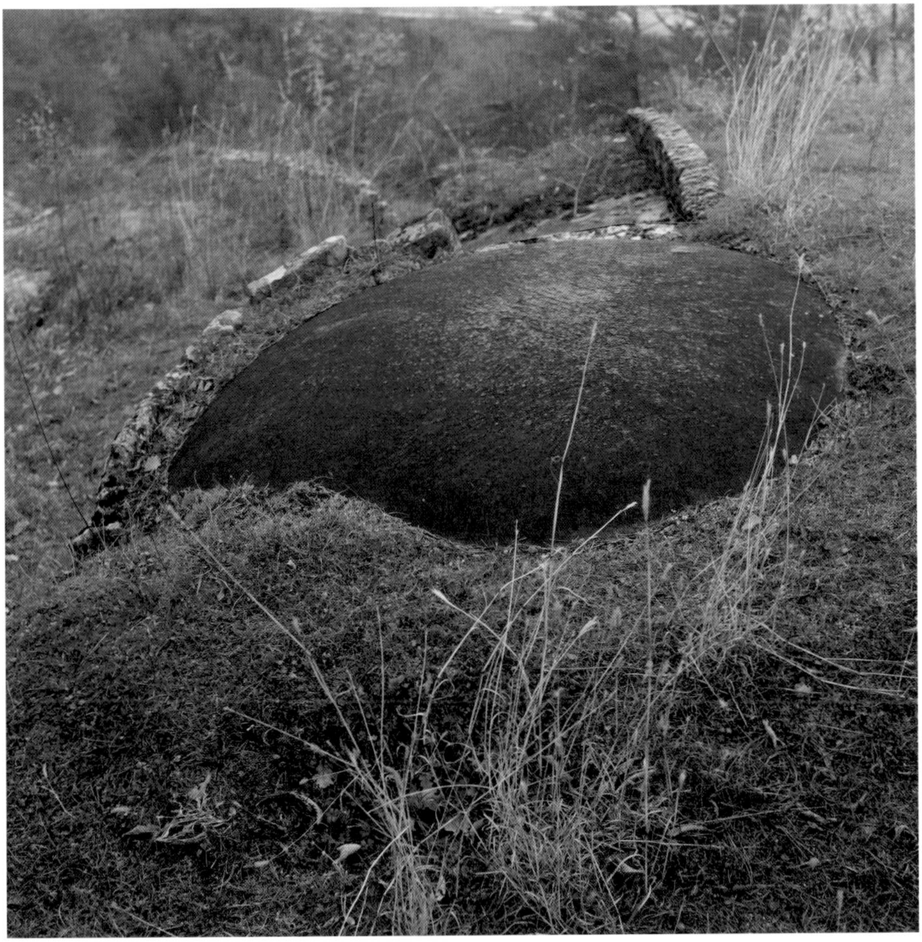

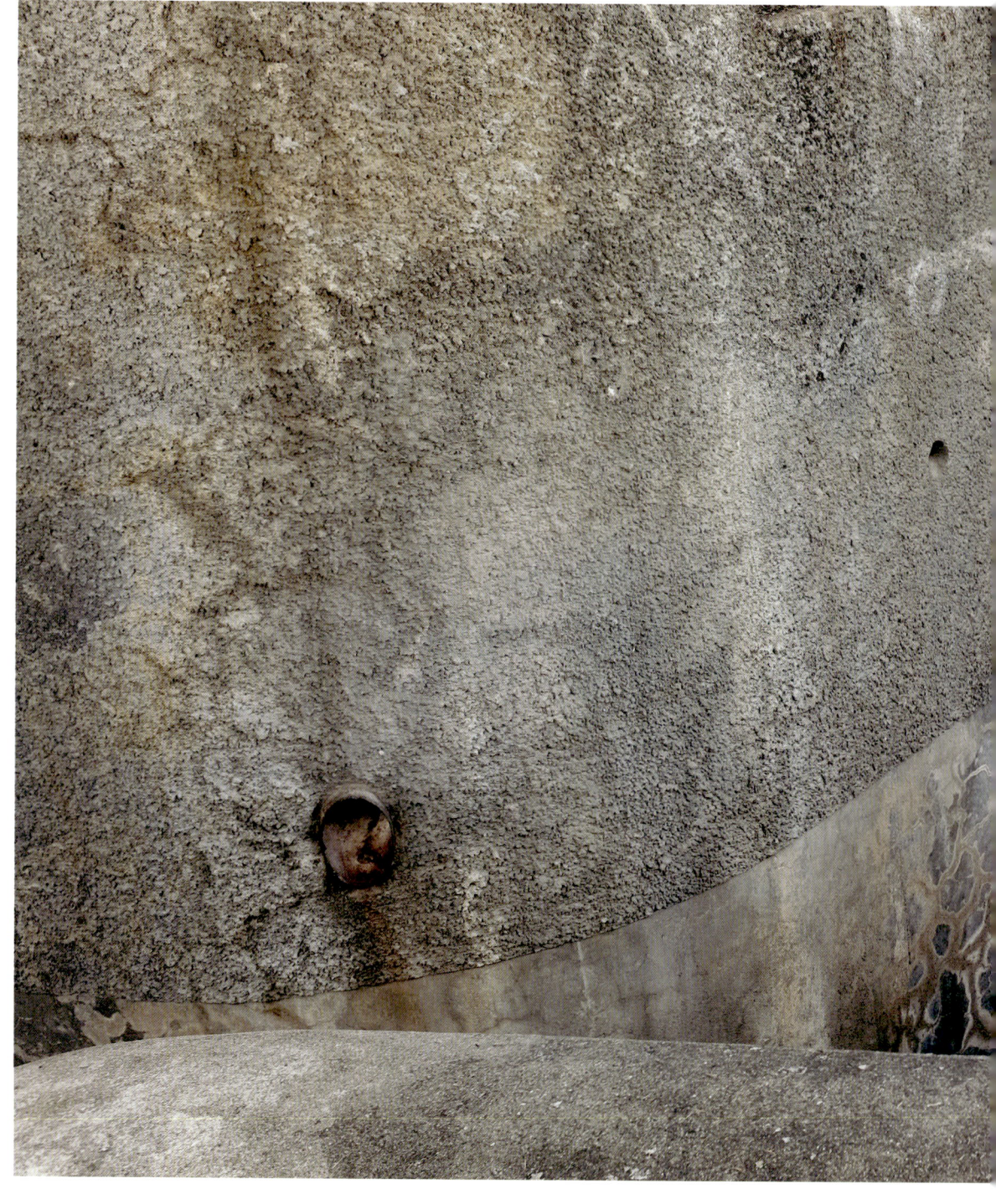

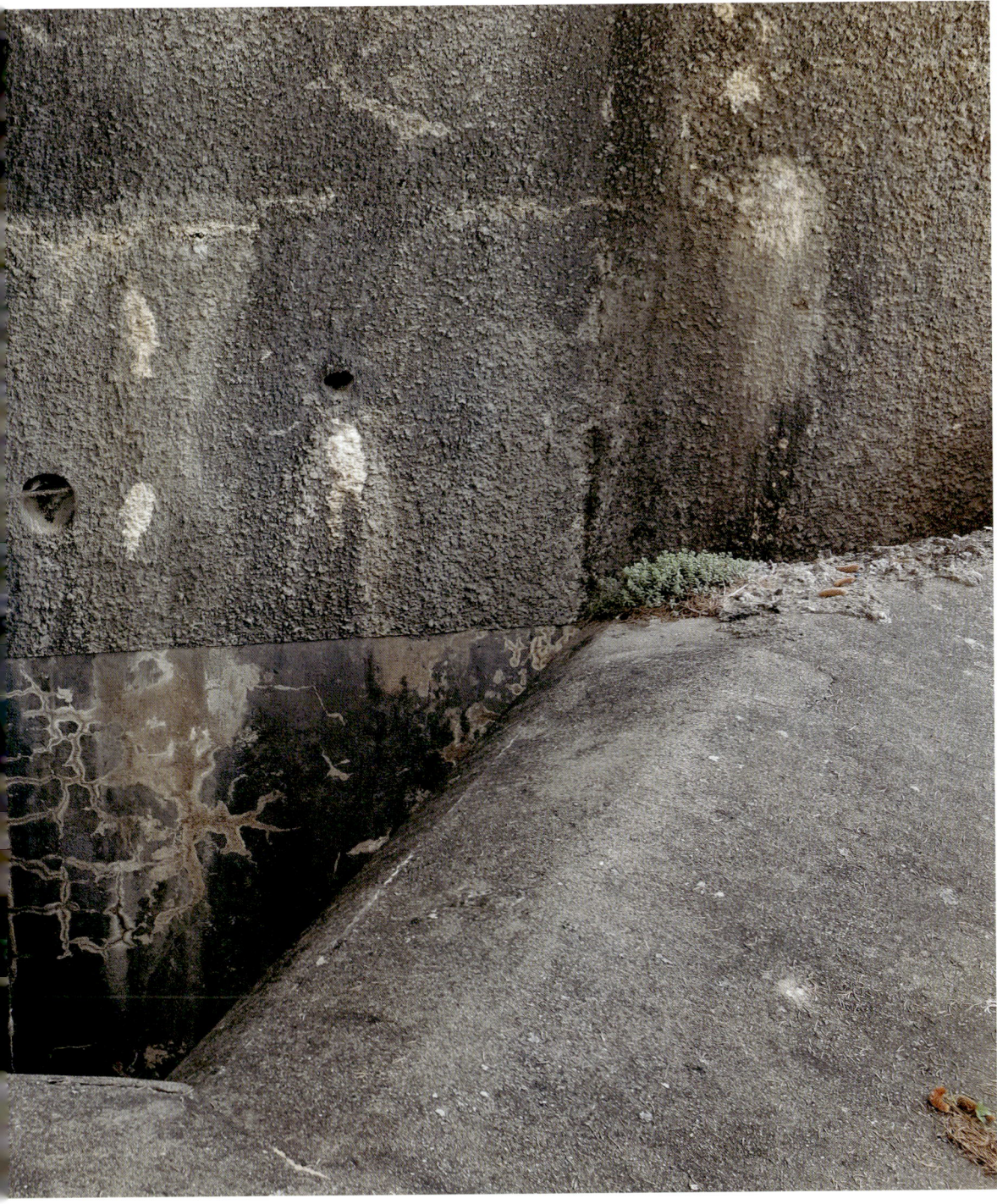

Ouvrage de Flaut
SFAM / Tinée-Vésubie
Seven officers, 296 men

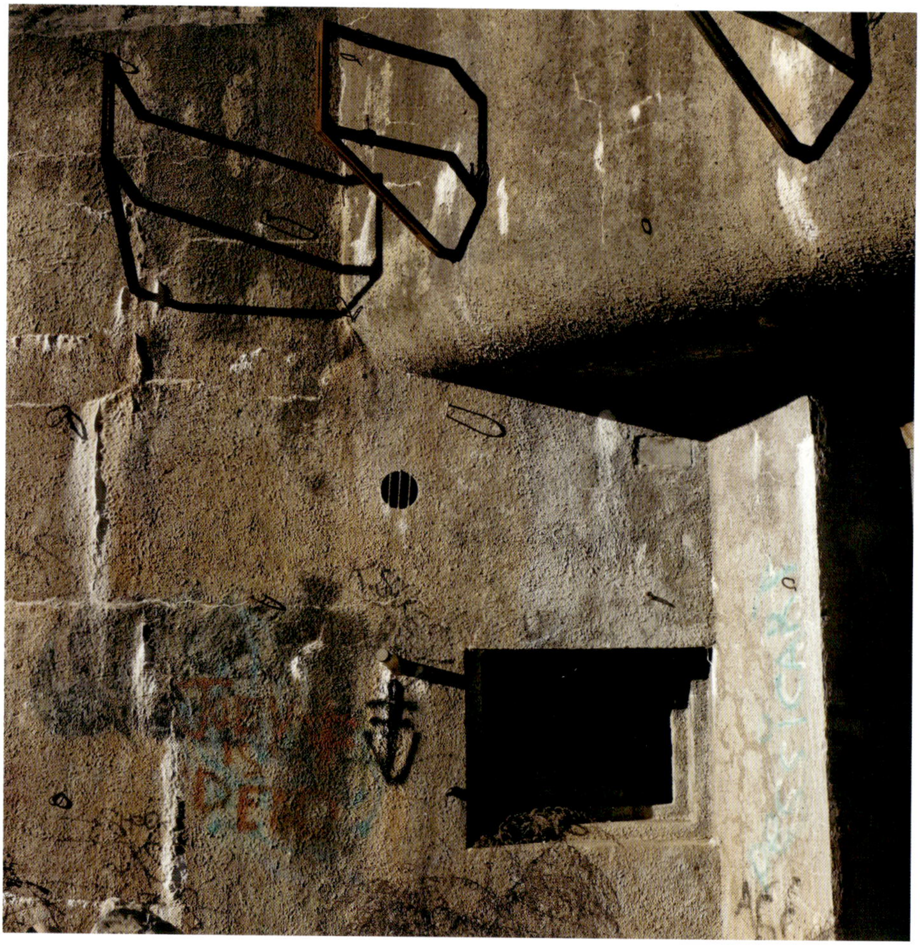

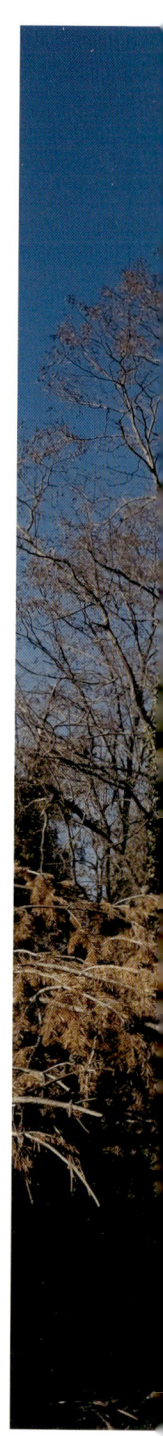

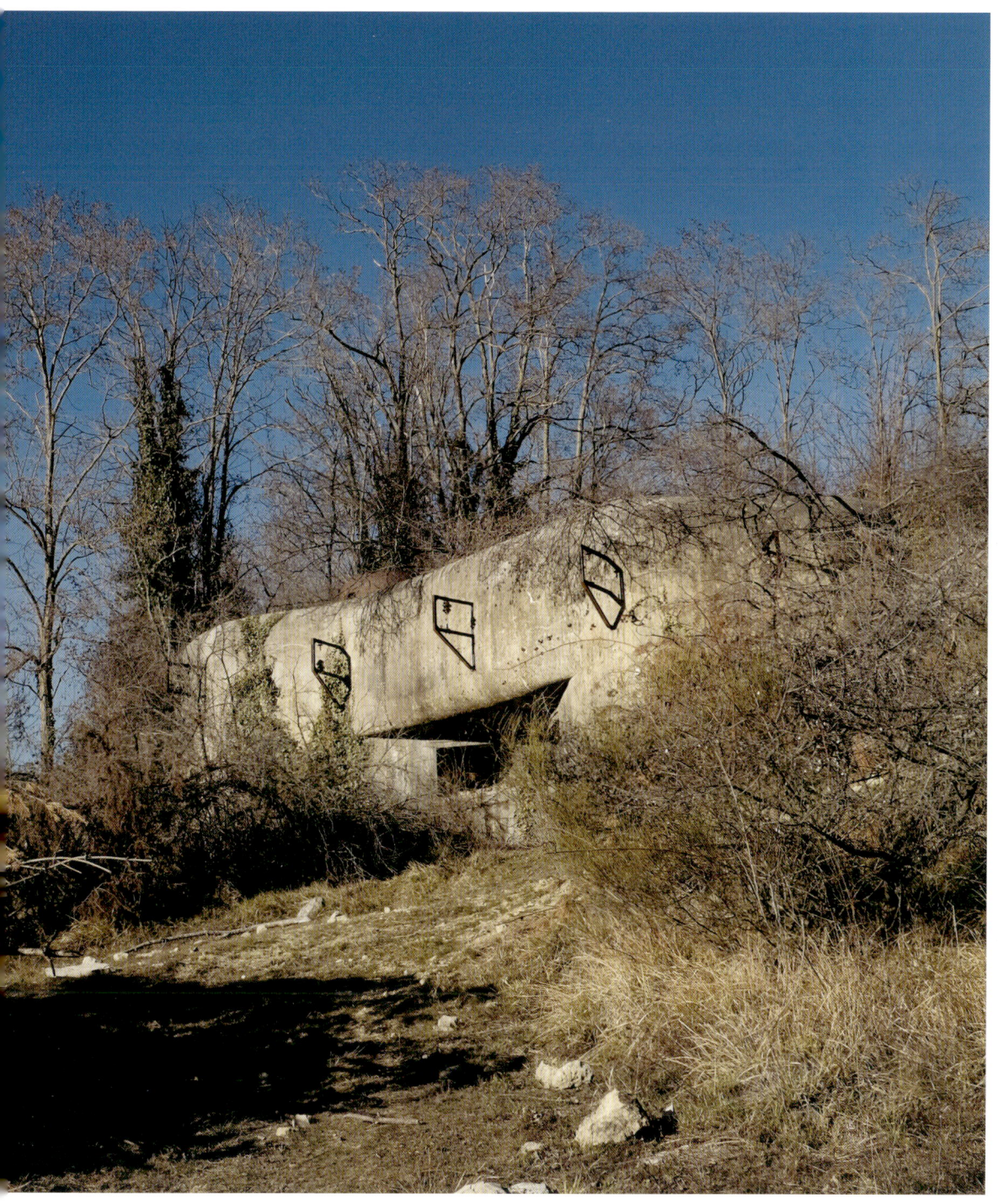

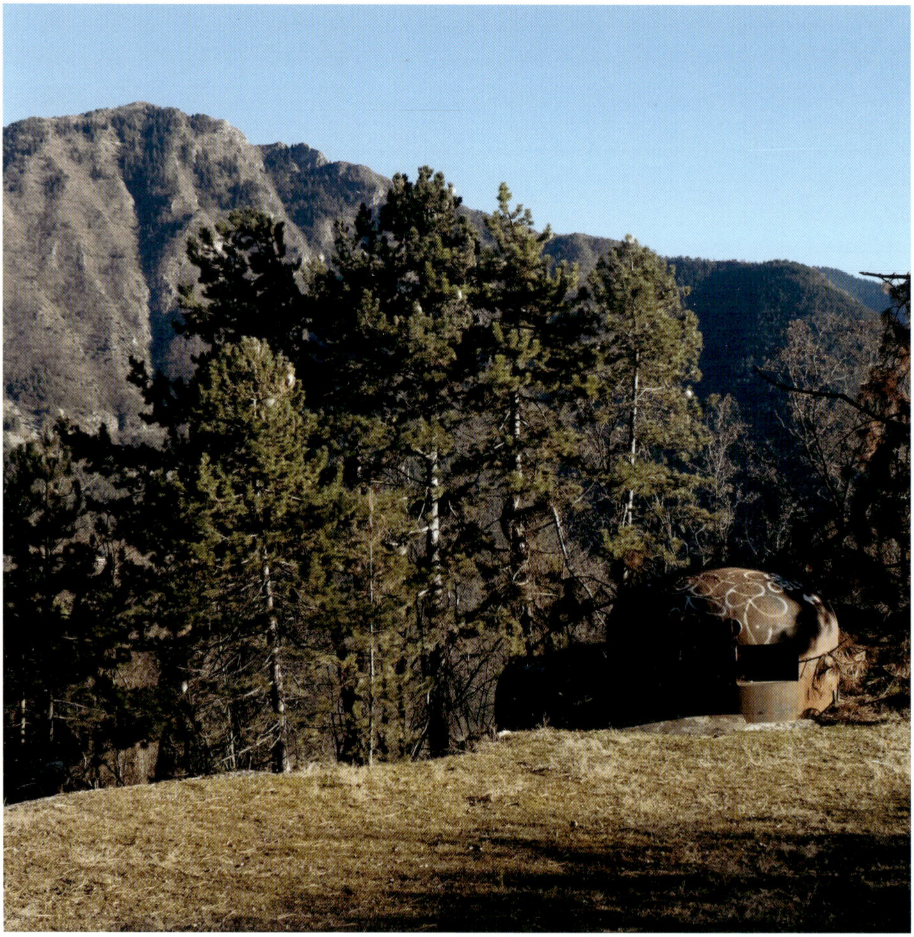

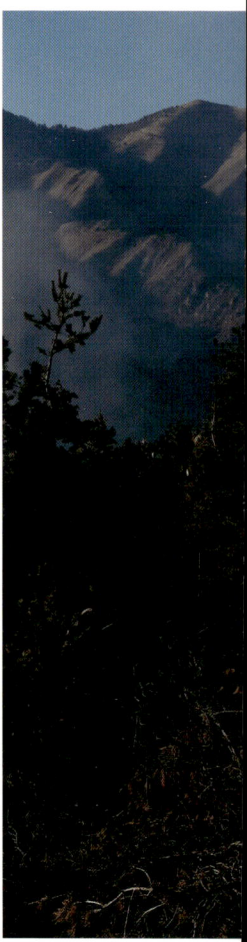

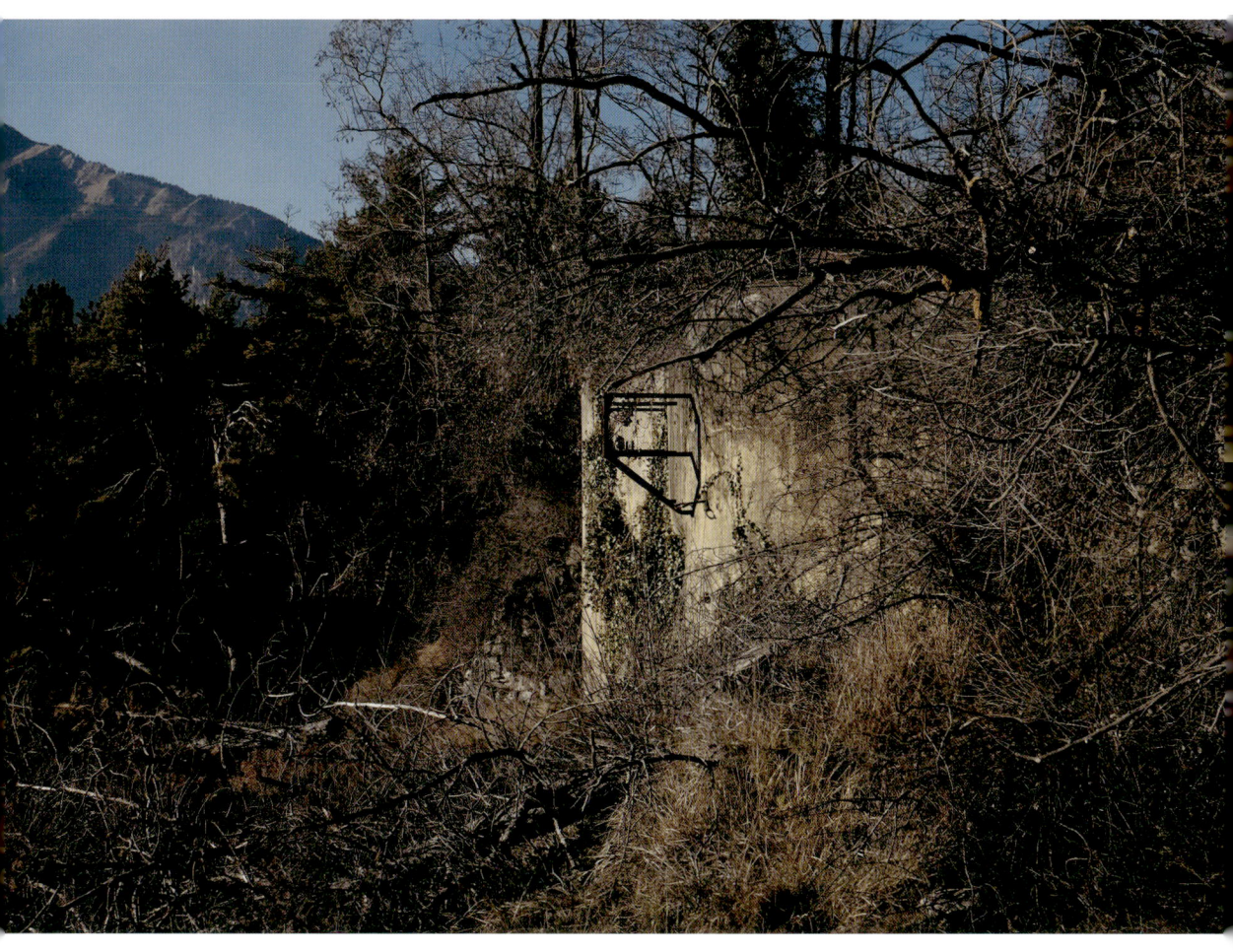

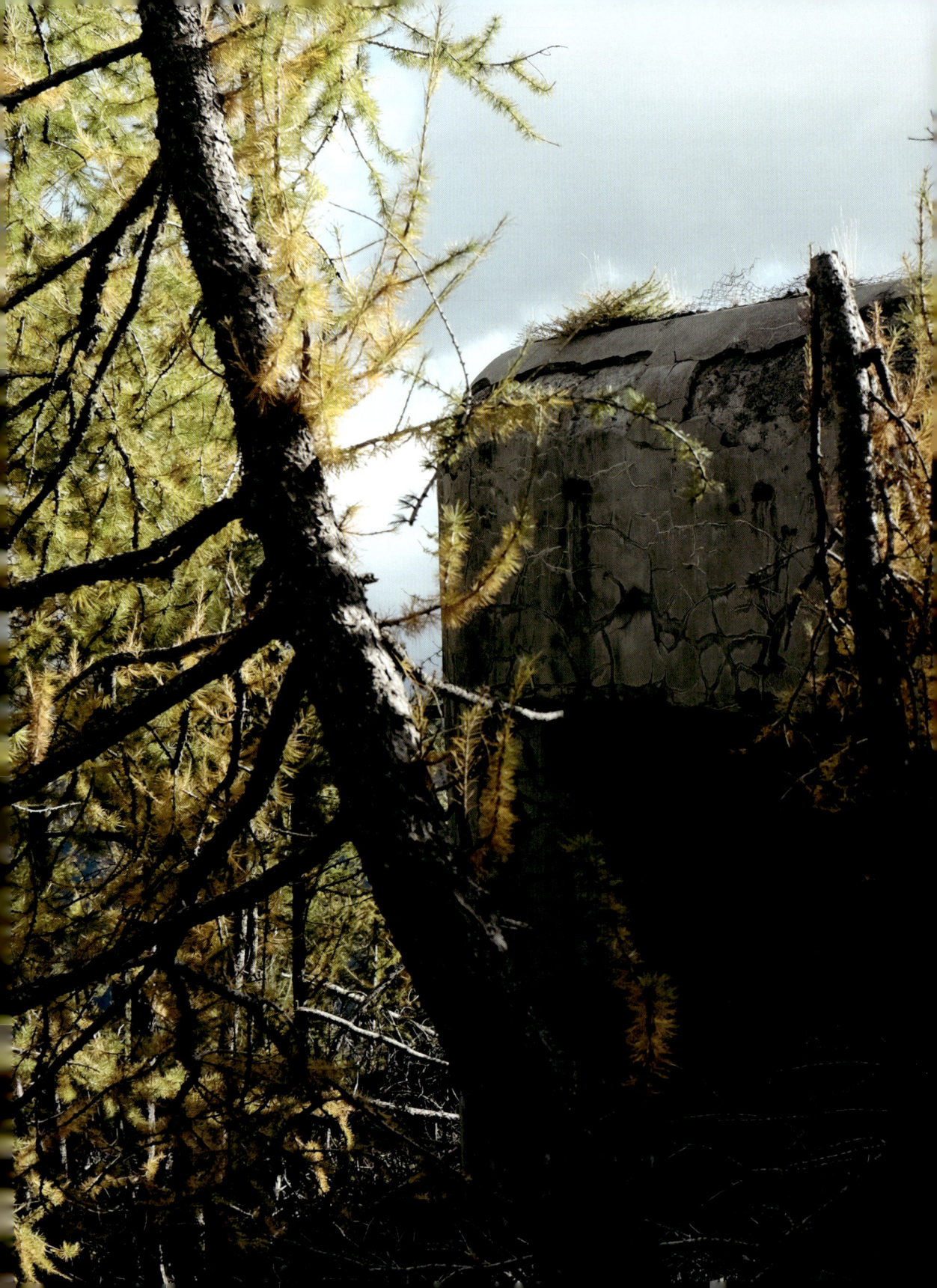

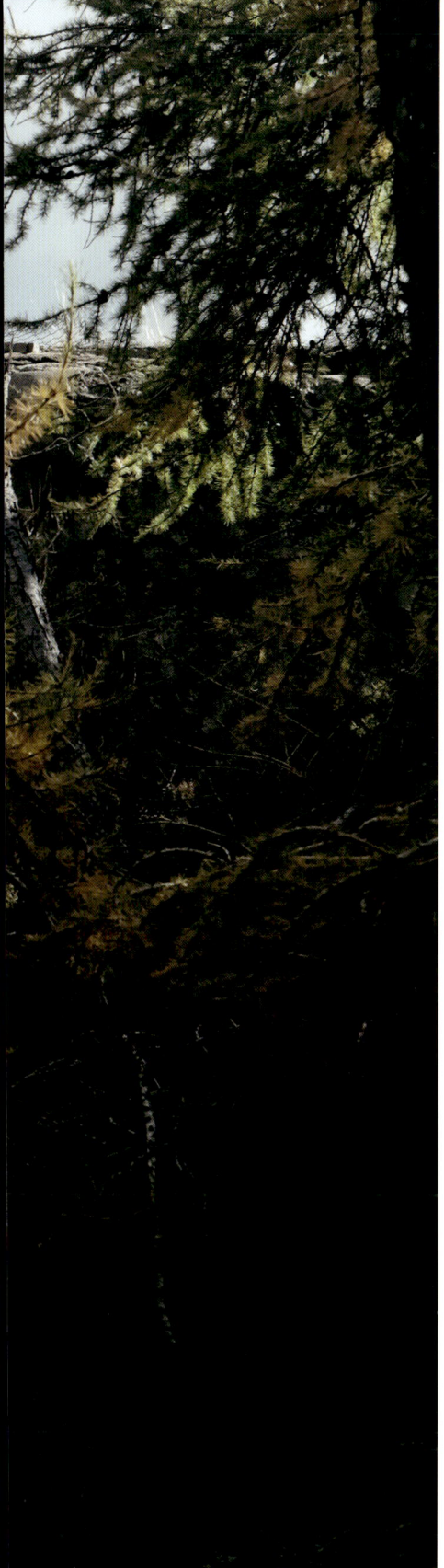

**Petit ouvrage
du Plan Caval**
SFAM / Authion
11 officers, 287 men

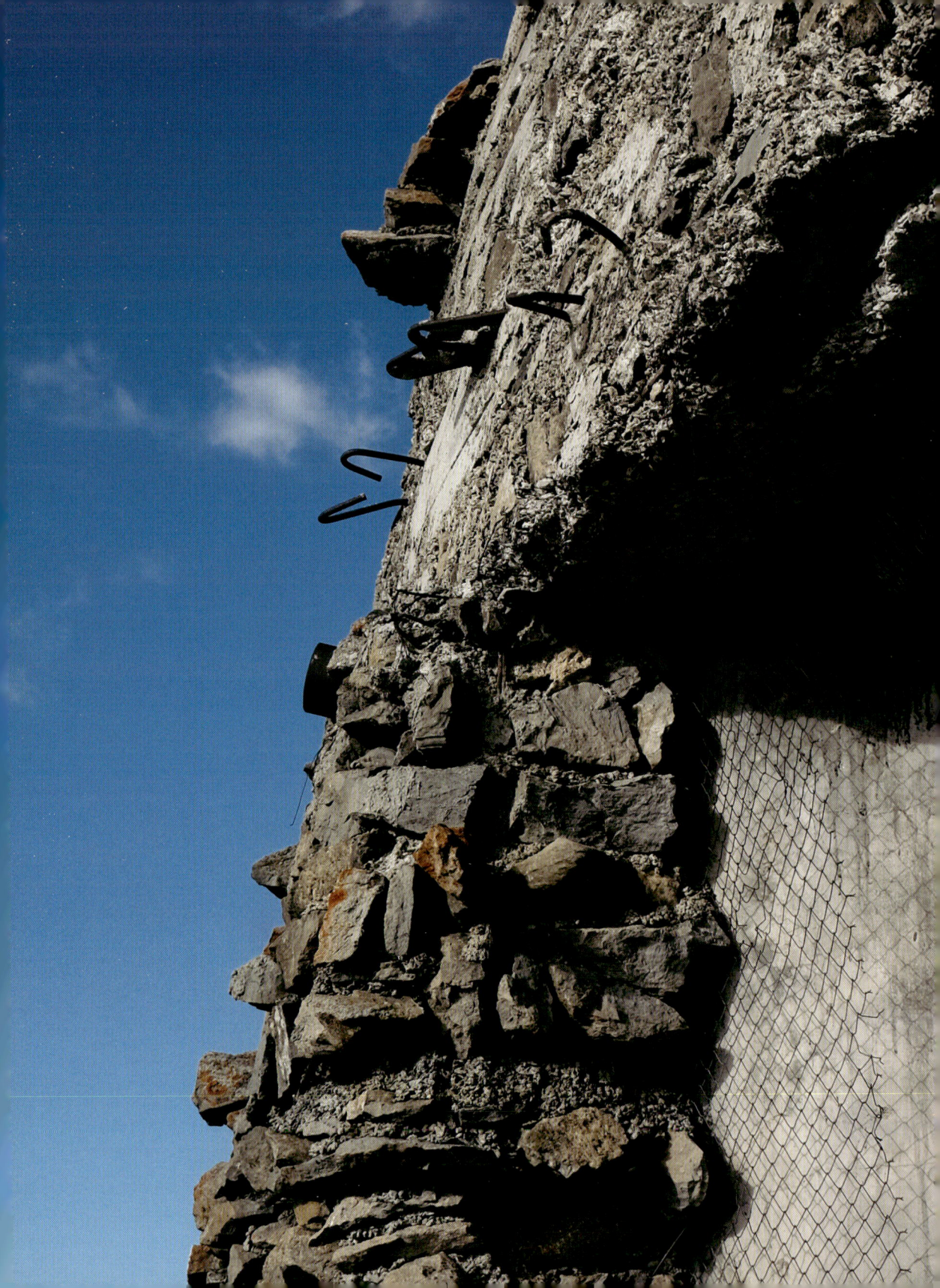

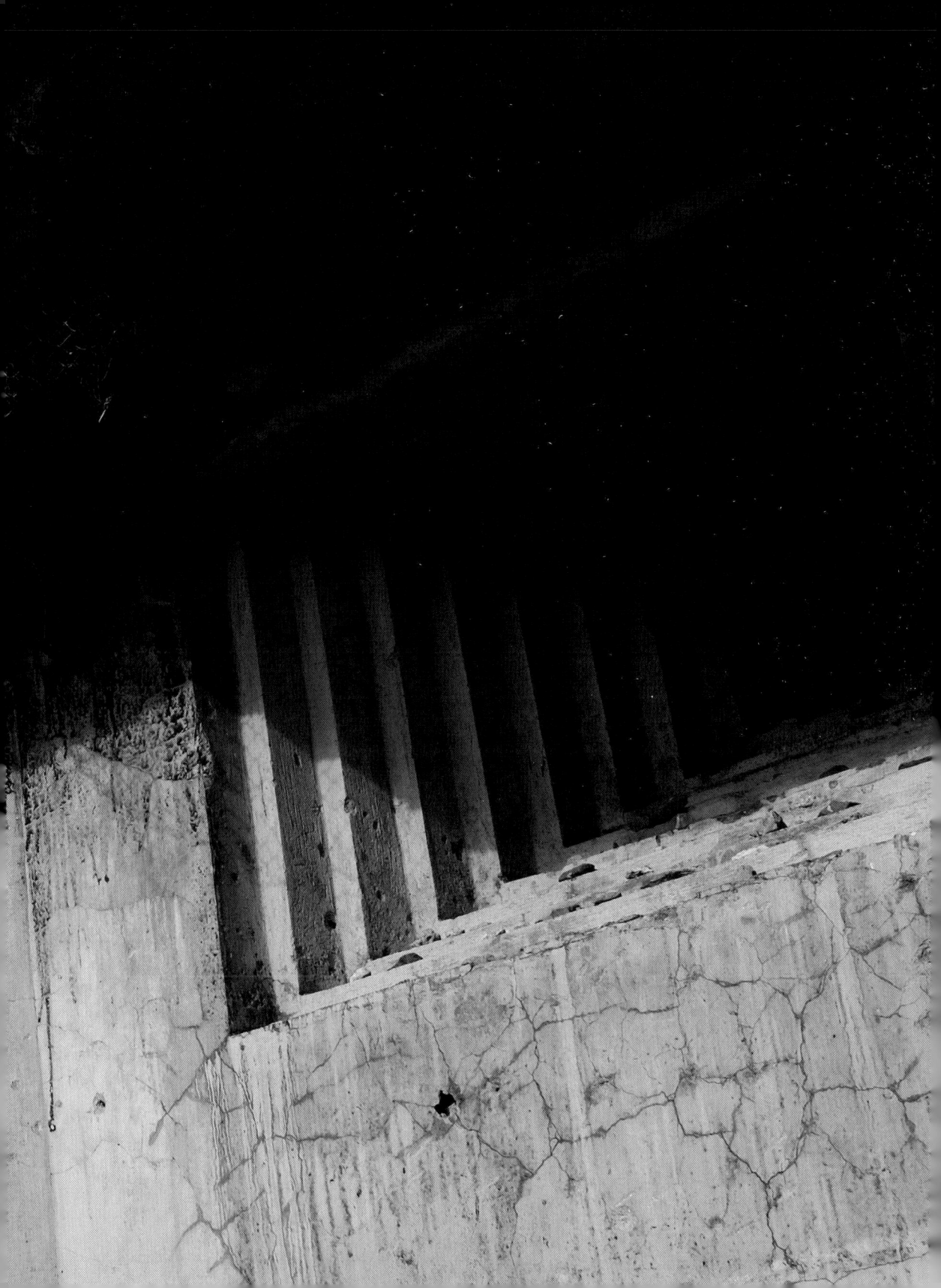

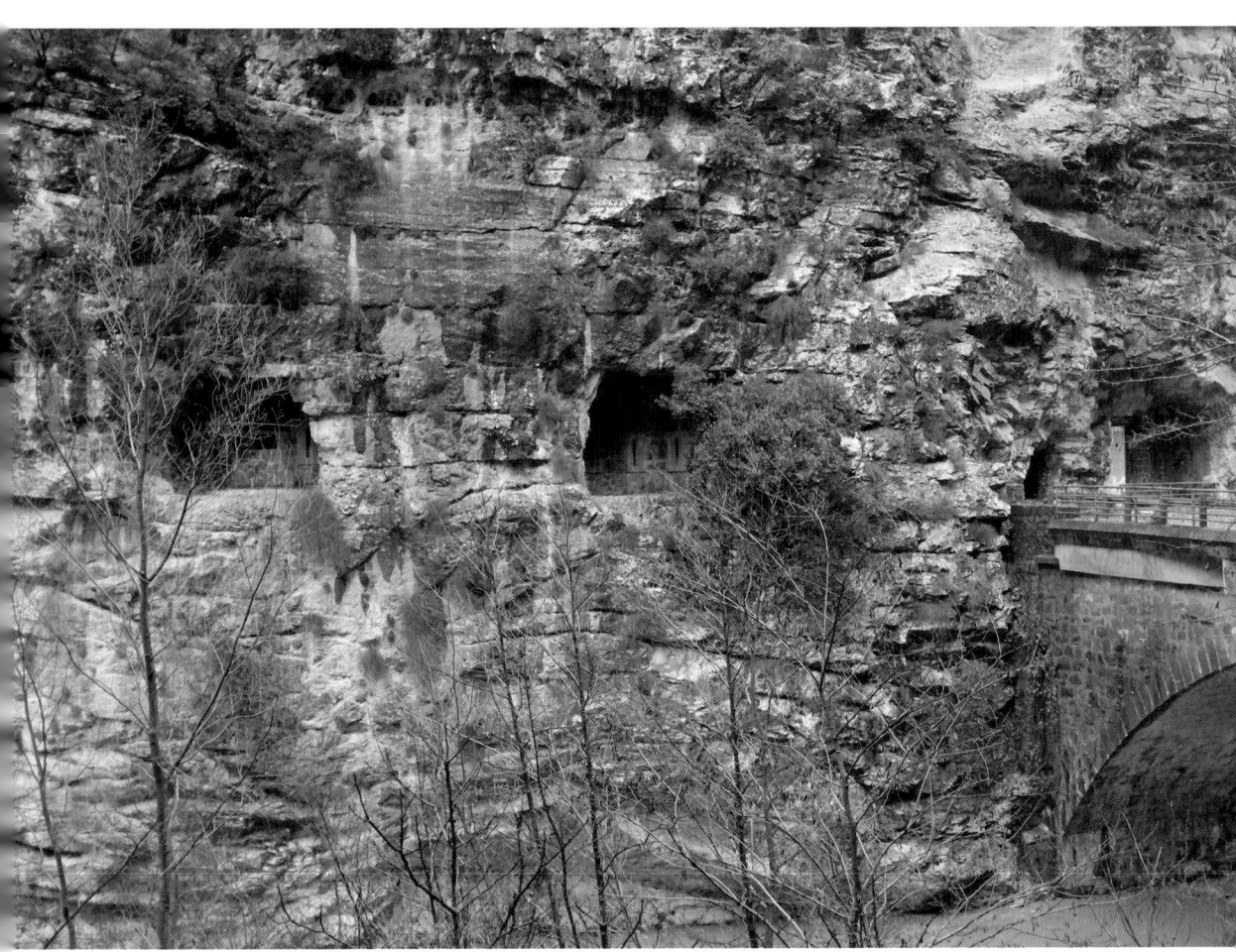

Chiuse de Bauma Negra
SFAM / Tinée-Vésubie

Built by the engineering corps between 1884 and 1887, this cave-fort controlled the bottom of the Tinée gorges, blocking progress towards Nice with heavy firepower and a retractable bridge. On the right bank, barracks; on the left bank, cannon positions.

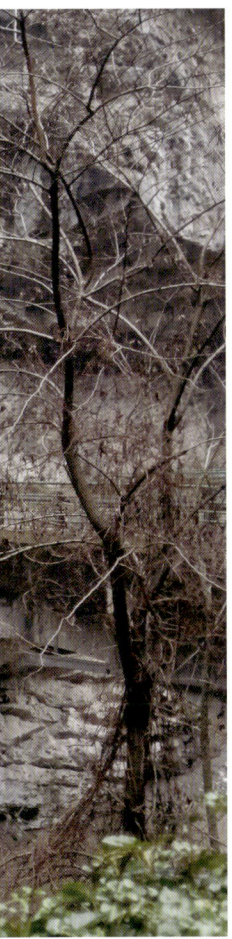

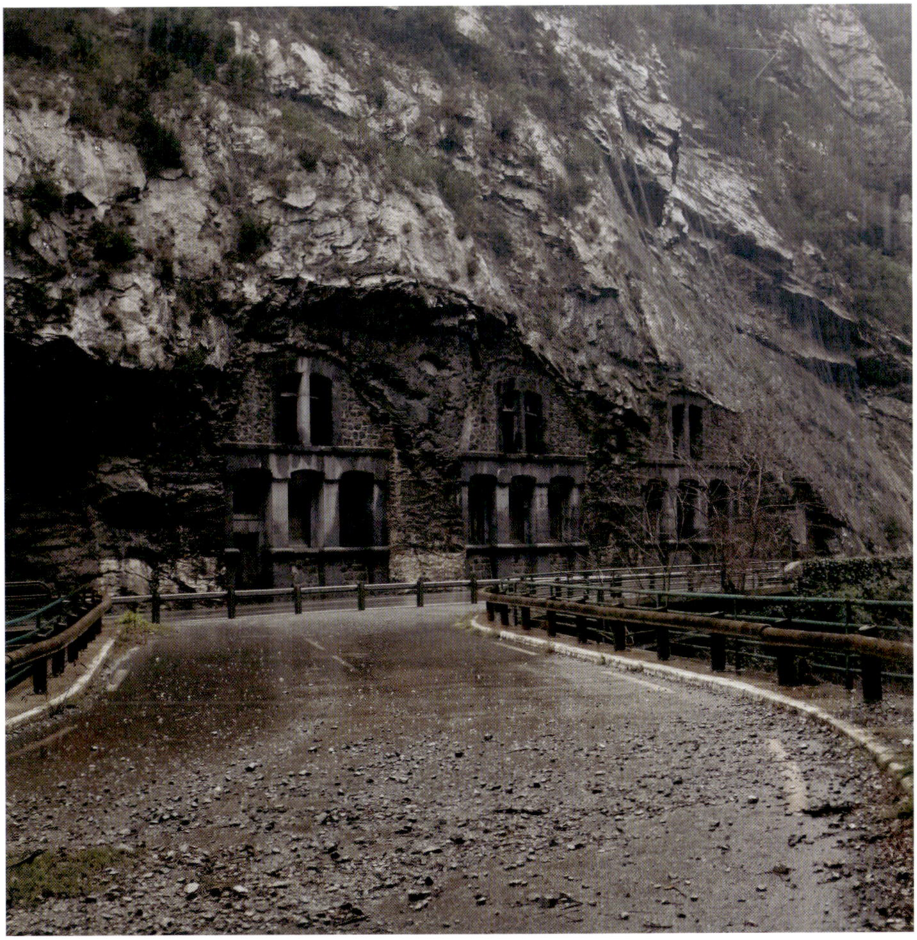

Gros ouvrage de Sainte Agnès
SFAM / Corniches
Eight officers, 310 men

Sainte Agnès played a part in the defence of Menton when the Italians attacked in 1940, firing some 1,500 times with 75mm and 135mm guns, and 81mm mortars.

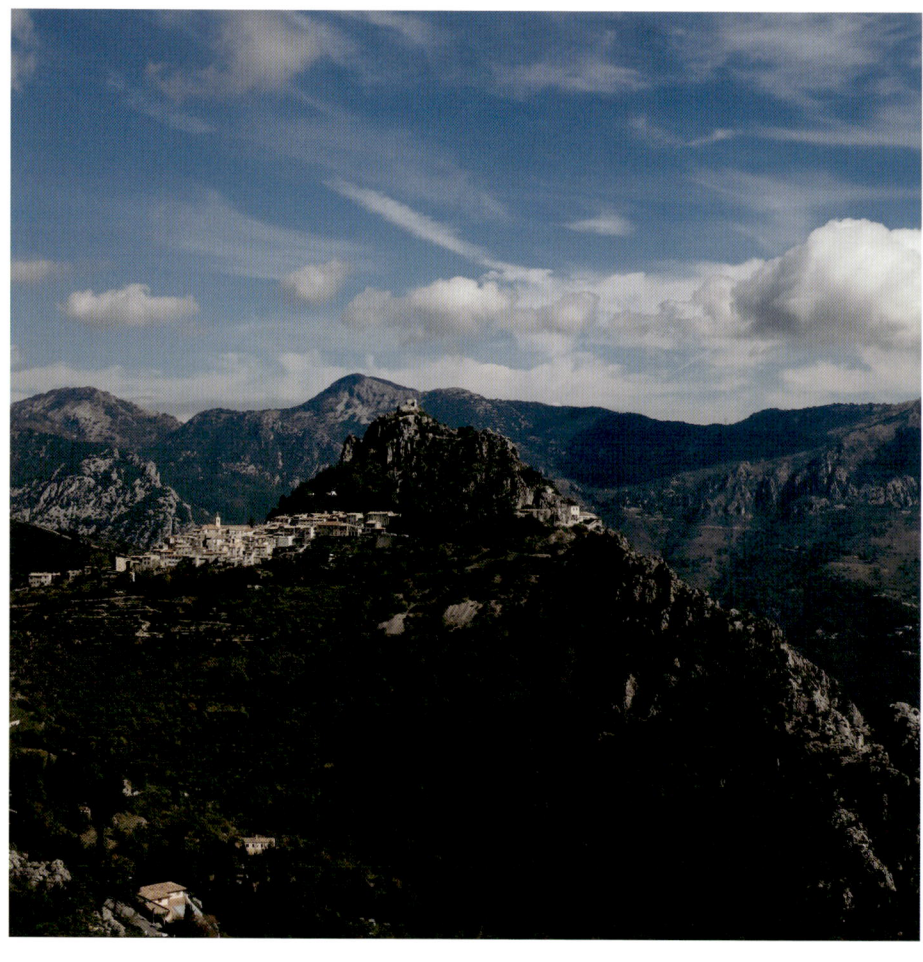

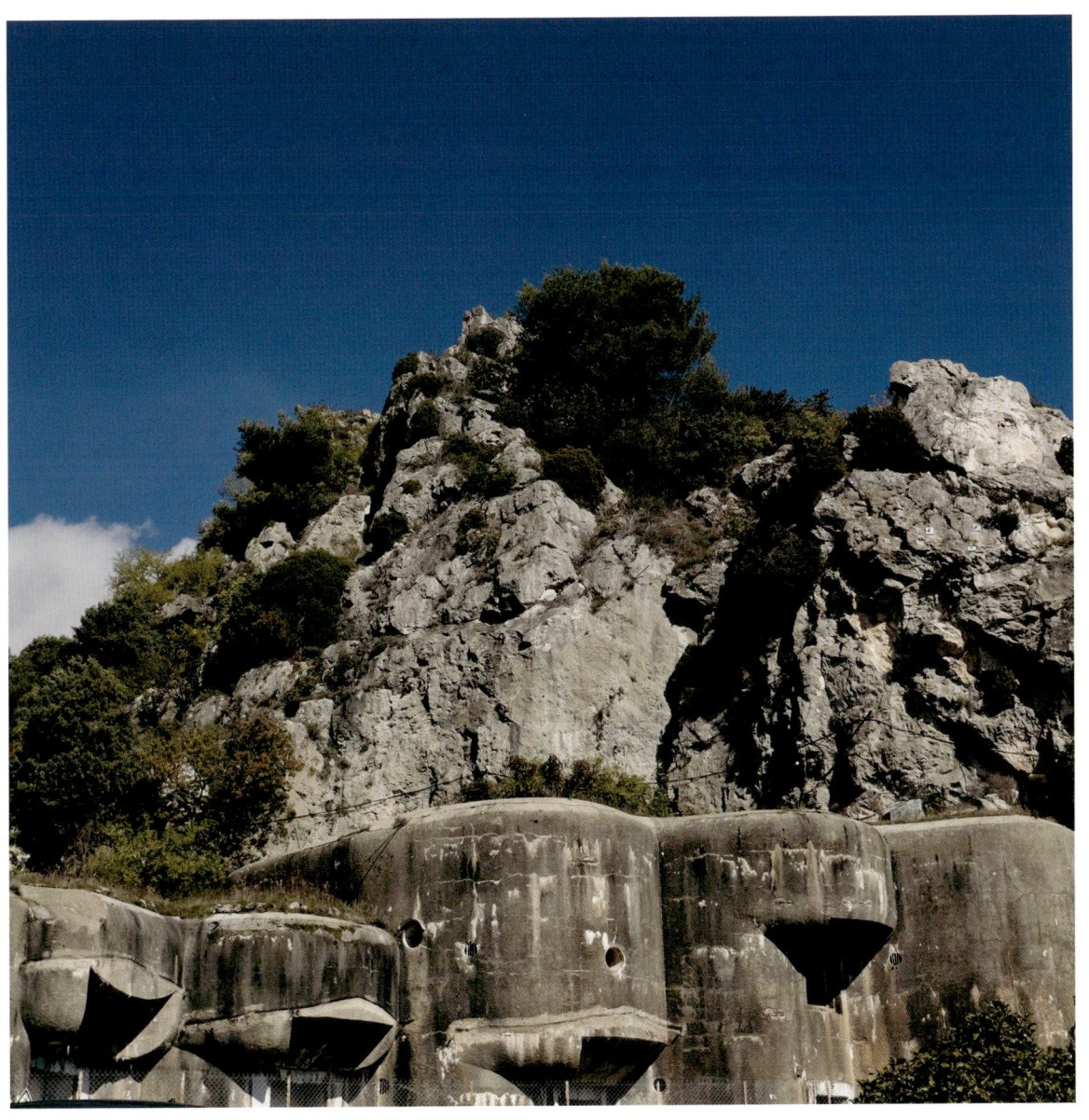

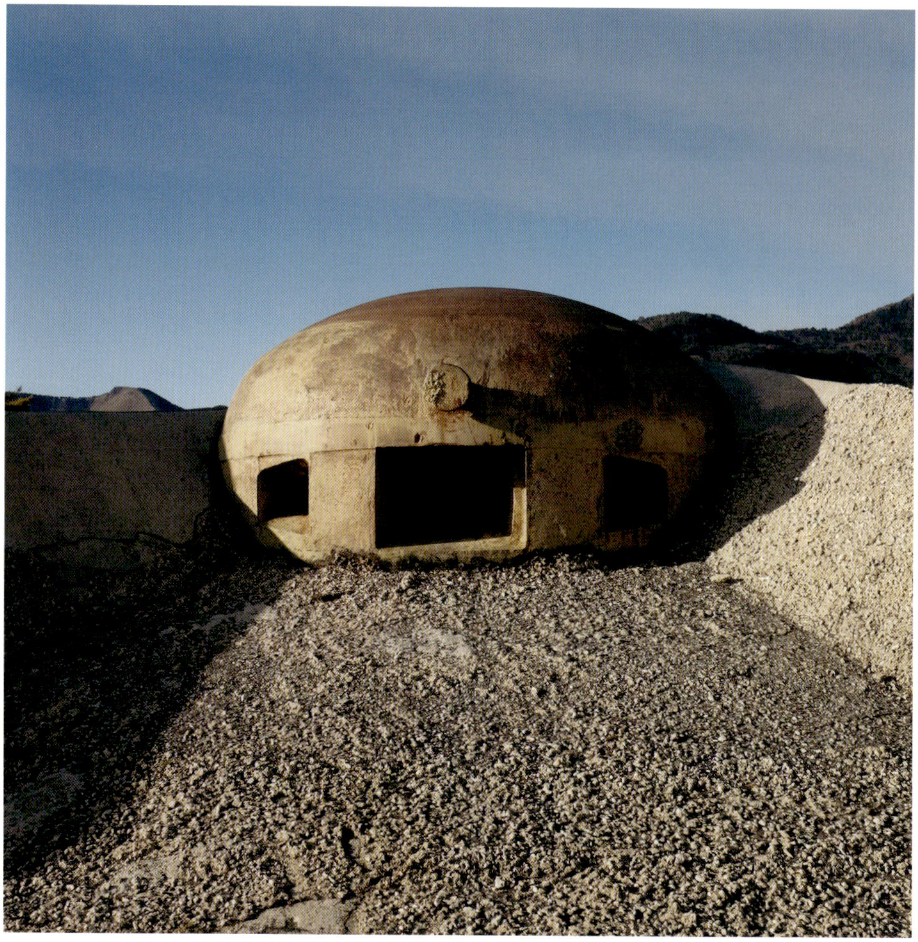

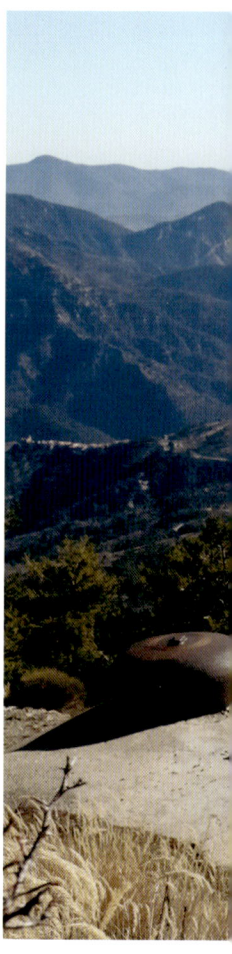

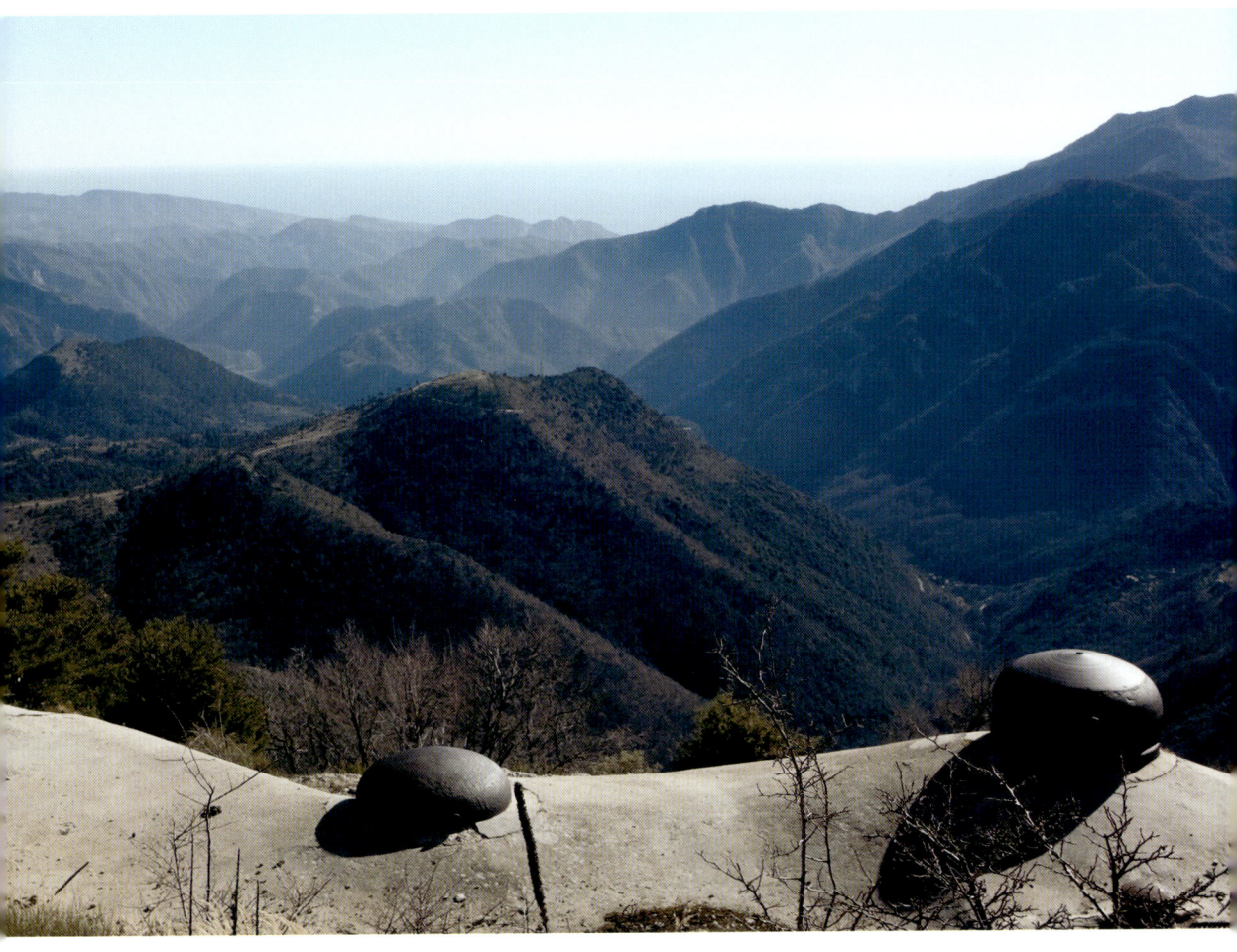

Above
Gros ouvrage de Monte Grosso
SFAM / Sospel
10 officers, 363 men

Following pages, left
La Redoute de la Pointe des Trois Communes
SFAM / Authion

The *redoute* was built in the 19th century and was the first fort in the Alps to be constructed from reinforced concrete. It was occupied by the Germans in 1944 and took heavy fire from De Gaulle's Free French Army in the Battle of the Authion in 1945.

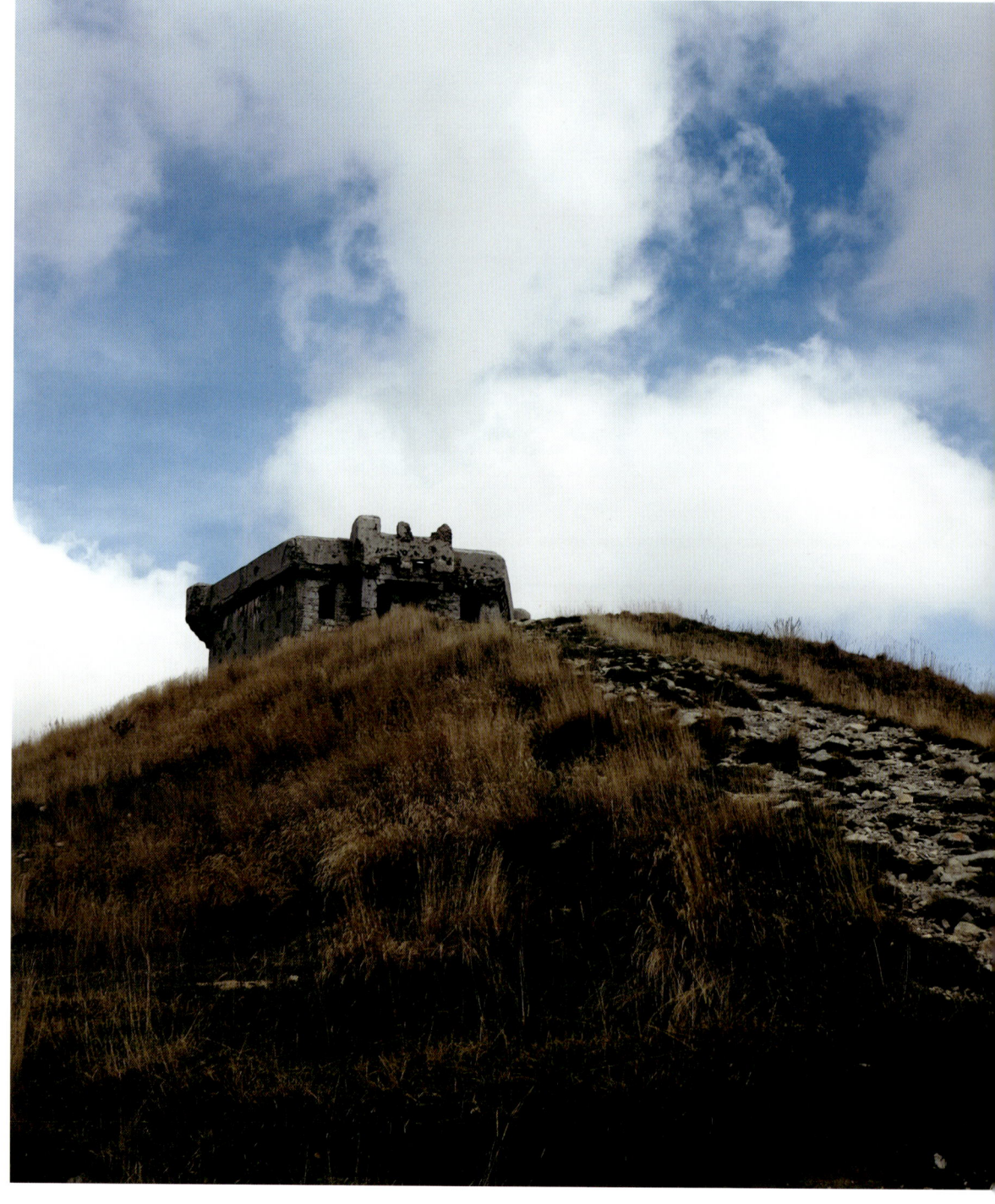

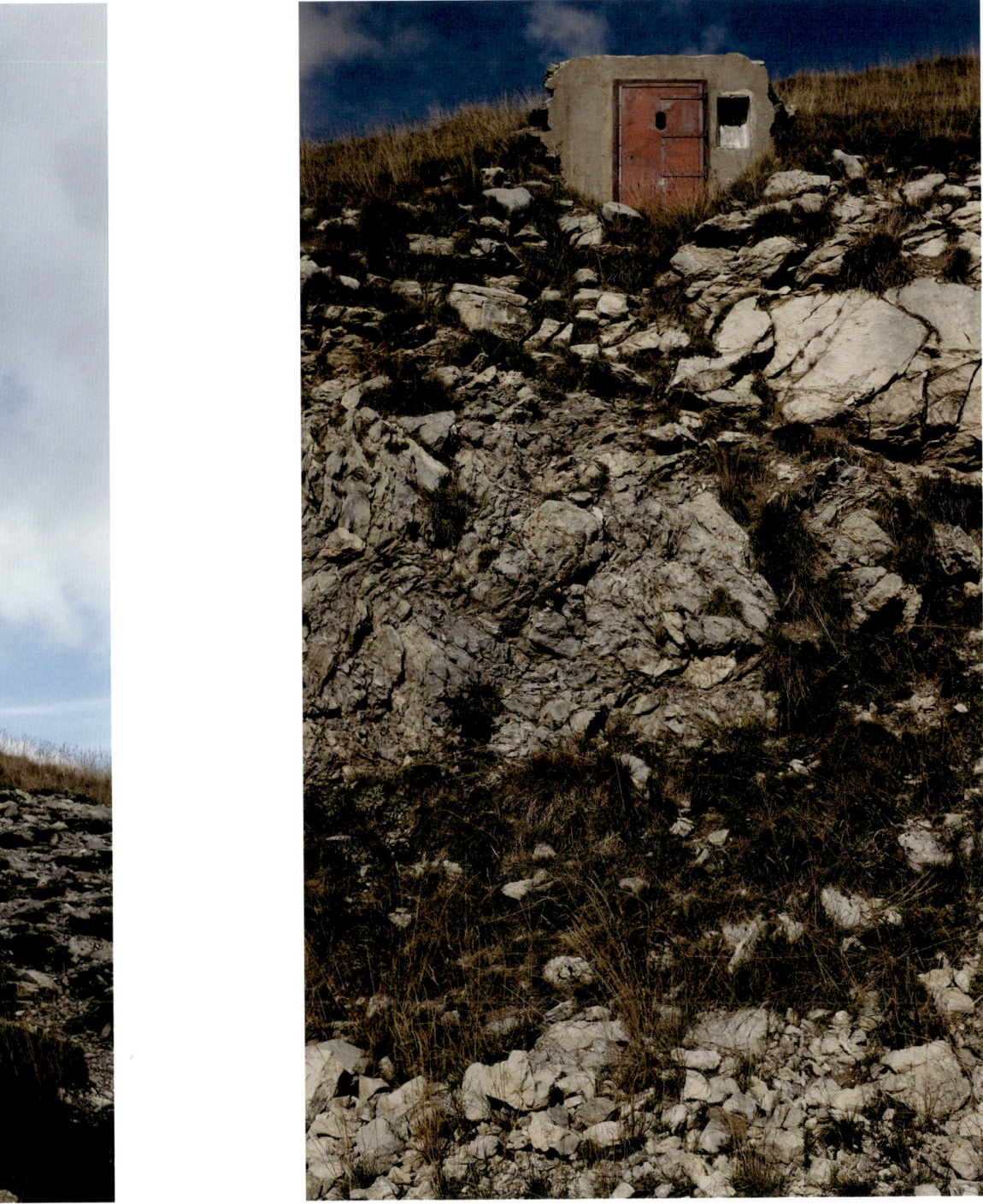

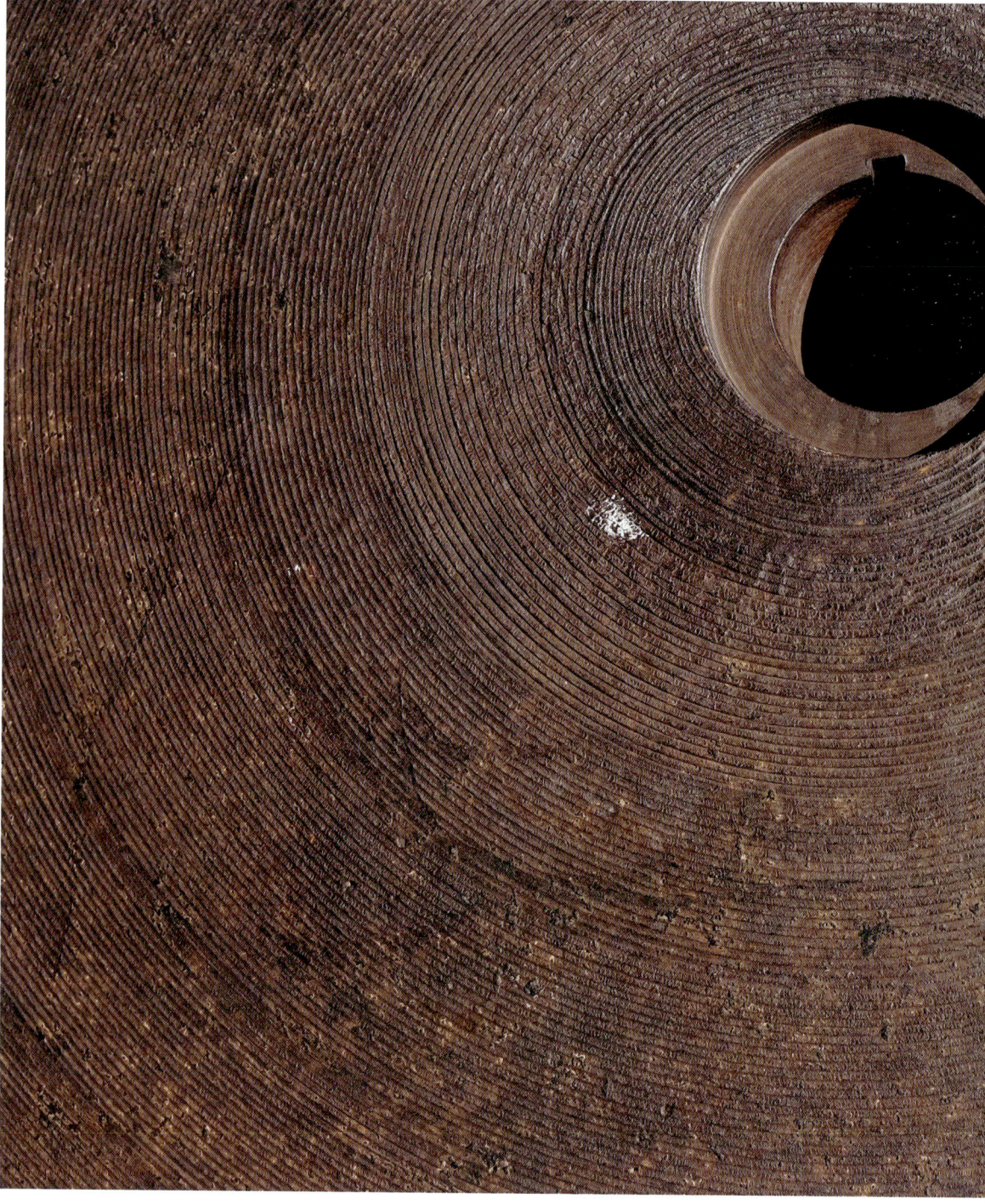

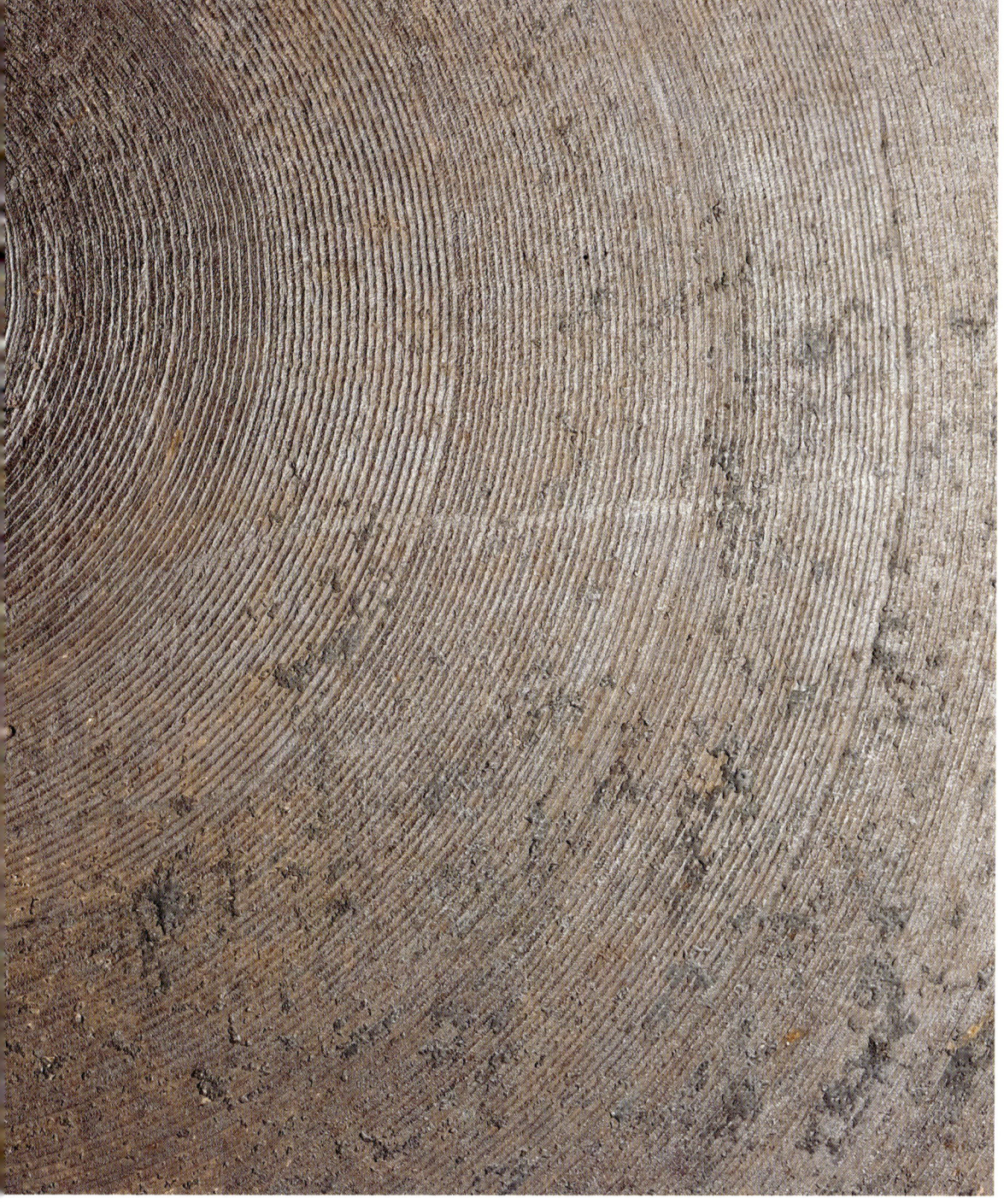

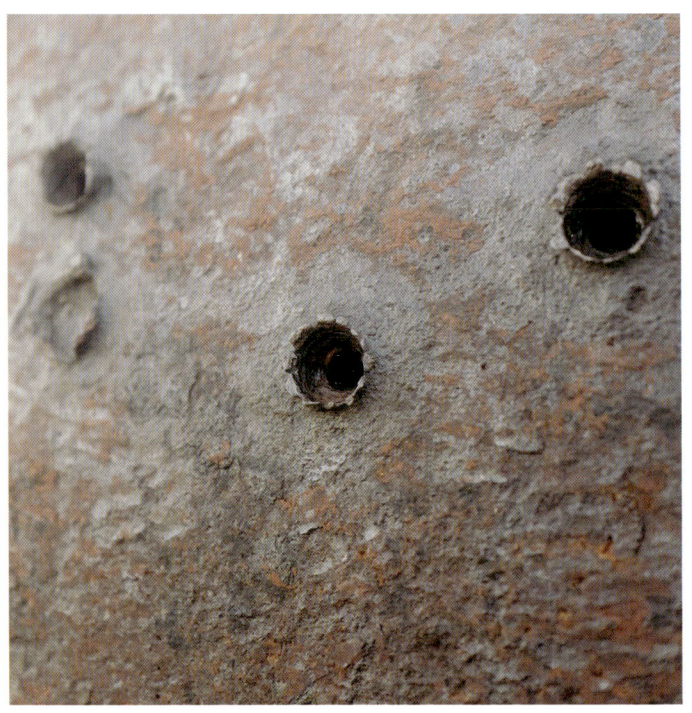
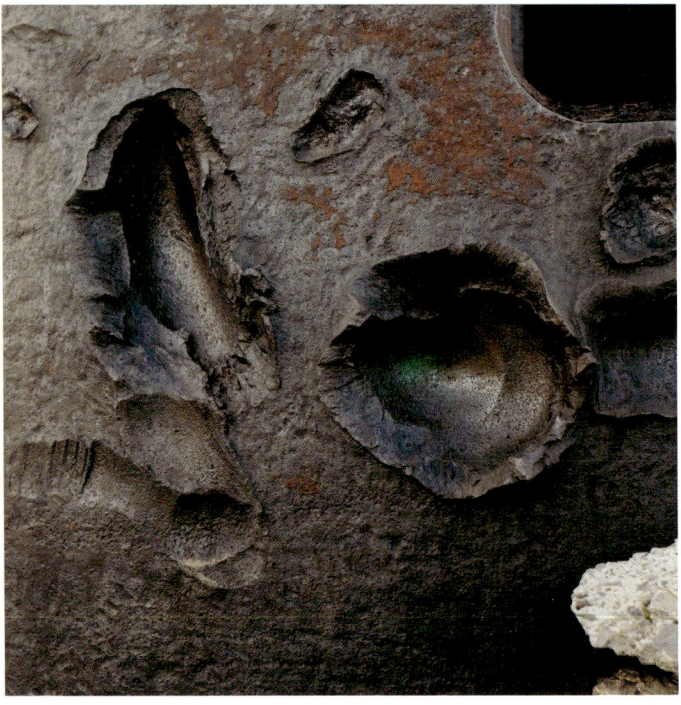
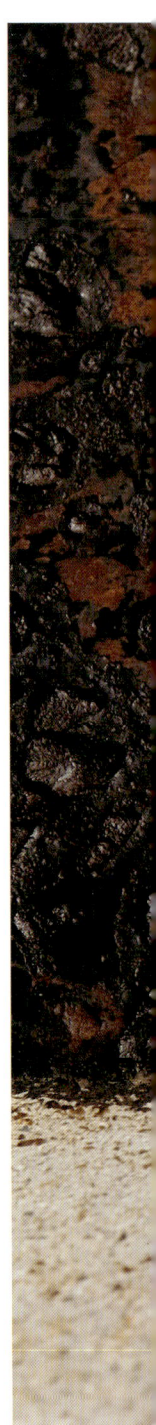

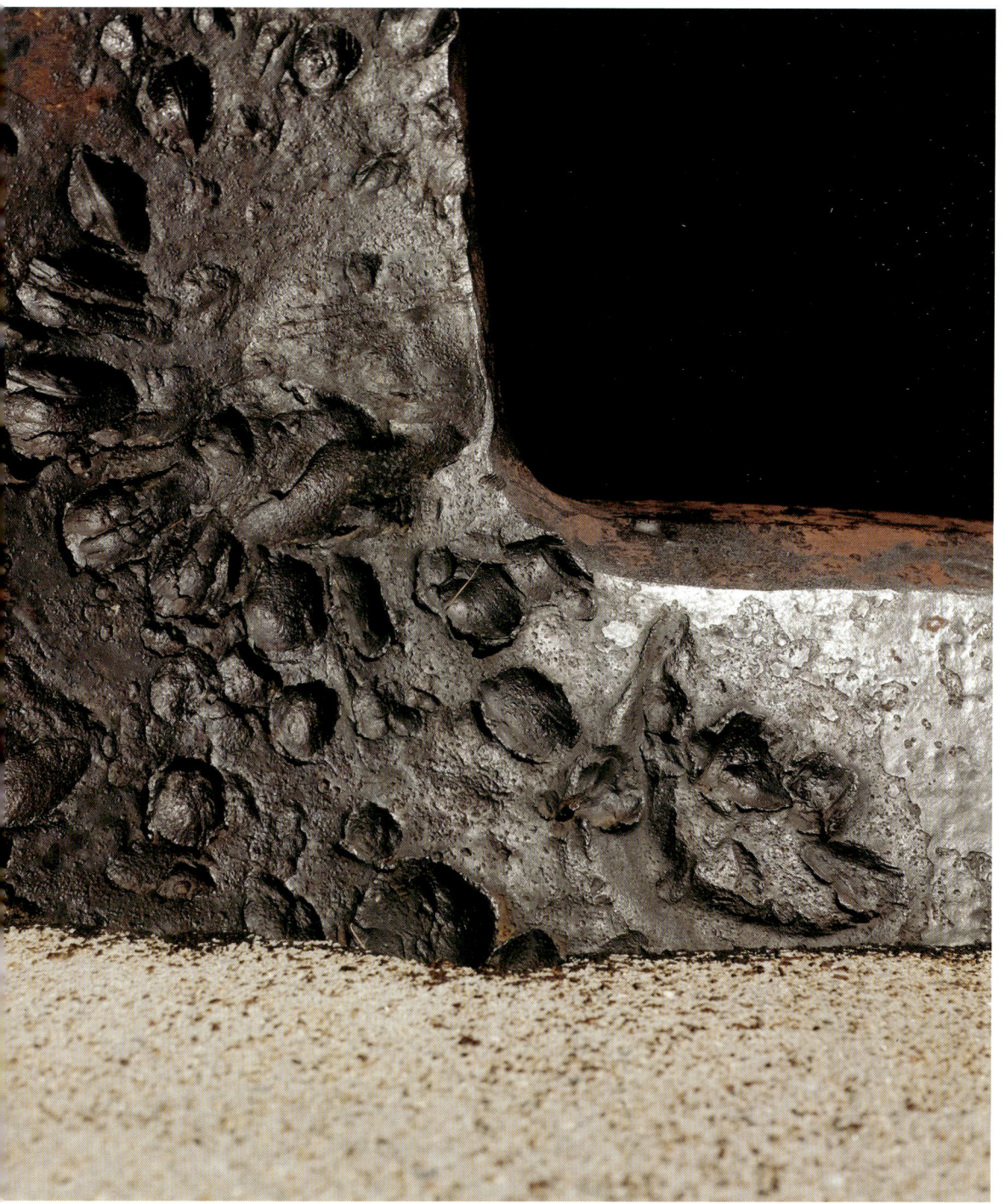

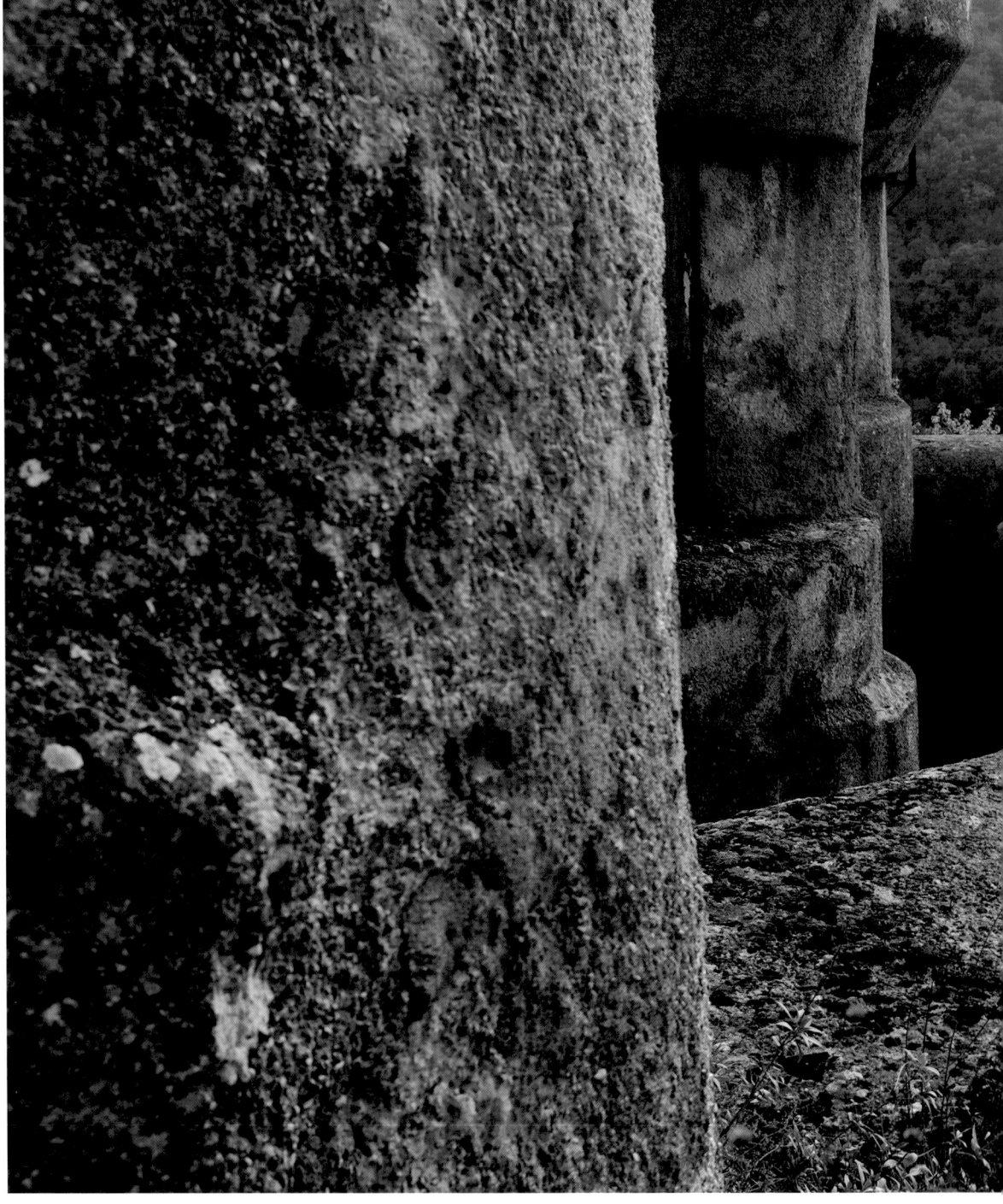

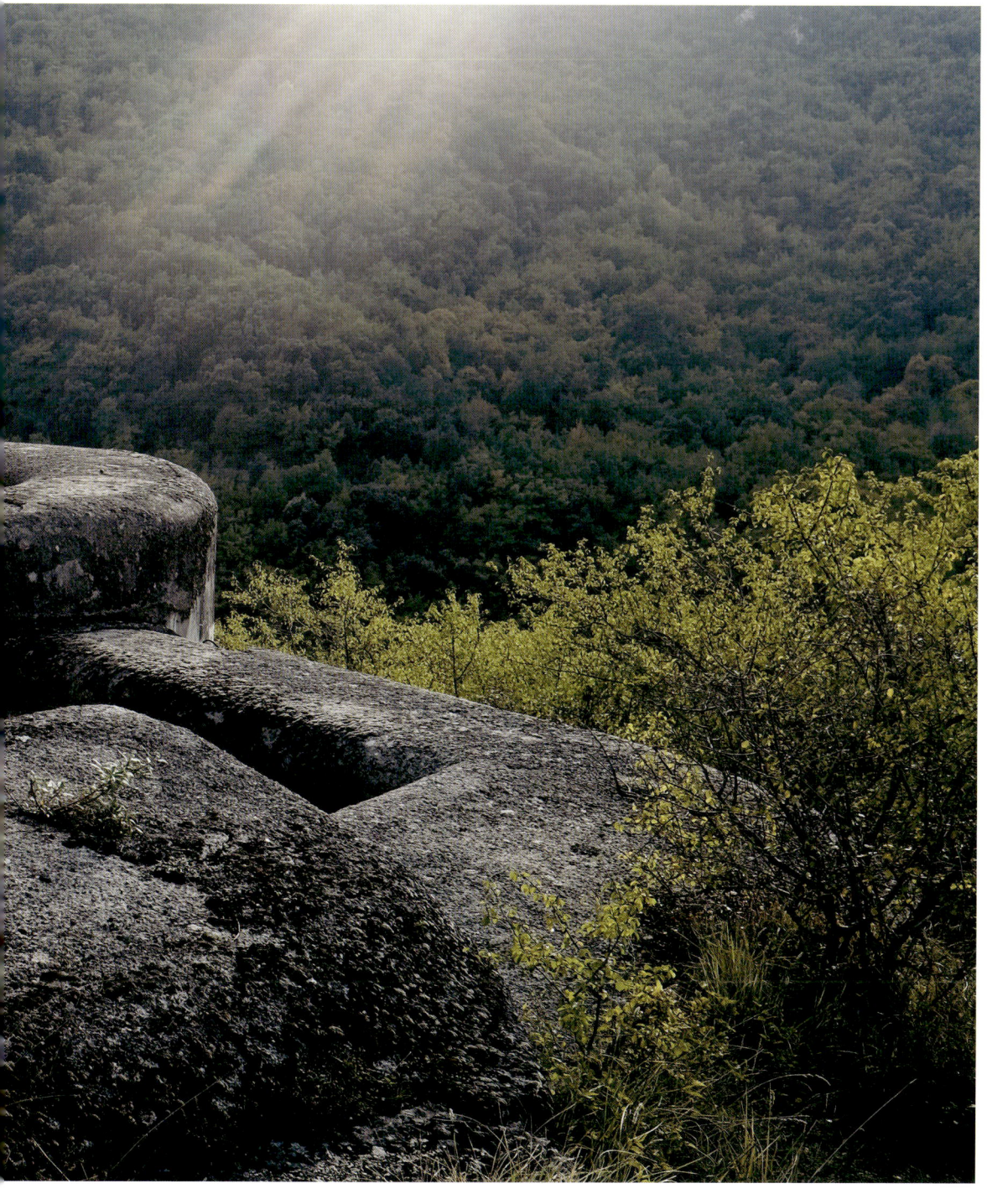

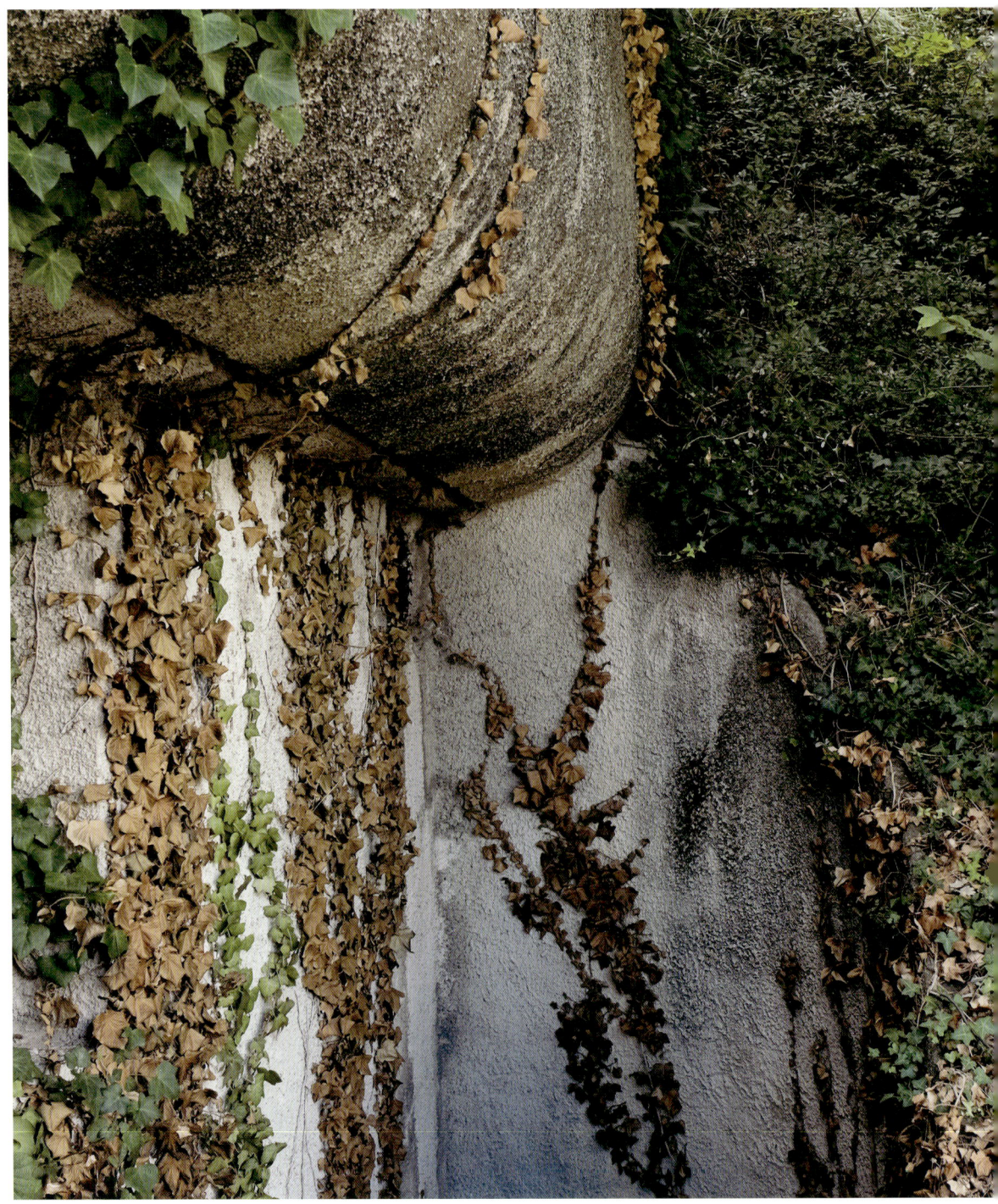

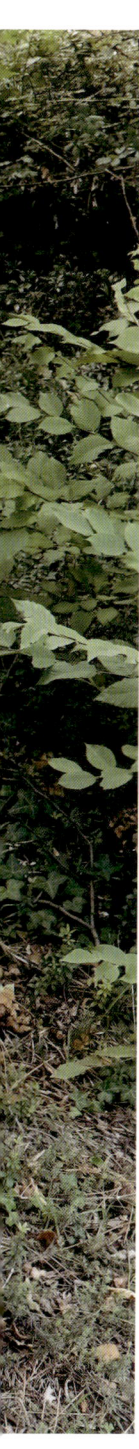

Gros ouvrage de Castillon
SFAM / Corniches
Seven officers, 337 men

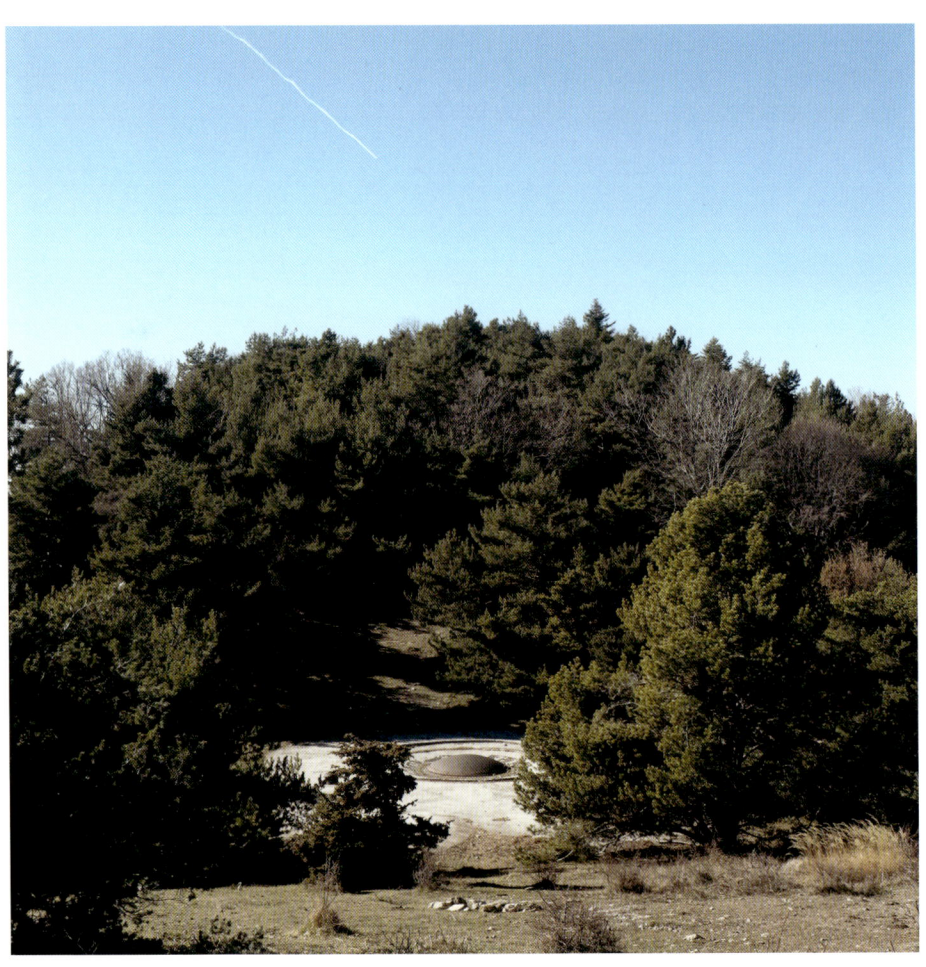

« Everybody will be entrenched in the next war. It will be a great war of entrenchments. The spade will be as indispensable to a soldier as his rifle. The first thing every man will have to do, if he cares for his life at all, will be to dig a hole in the ground. »

Is War Now Impossible?
Jan Bloch, 1901

Bunker Research

The Second World War fortifications of the Alpes-Maritimes

This is the story of an obsession; or rather, two obsessions running like parallel tracks over the course of more than a hundred years.

The most recent, that of a cyclist exploring the roads of the Alpes-Maritimes, the high mountains behind the Côte d'Azur, who sees the concrete fortifications guarding this, one of the most remote places in France, and thinks, 'Why?'

And then the originating obsession: the French national anxiety about the threat of an Italian invasion, a recurring paranoia with visible traces dating from the late 19th century and the years leading up to the Second World War.

These ruins of Third Republic forts and 20th-century bunkers catch the eye and exercise the imagination – squat cubes, soft geometric curves and elegant lines that both occupy and disappear into their surroundings, sitting in harmony and discord with the mountains around them.

Sited according to strategic military objectives in a pattern that is at first glance incomprehensible, they are, in an era of cooperation rather than threat, forlorn and purposeless. Marooned in some of

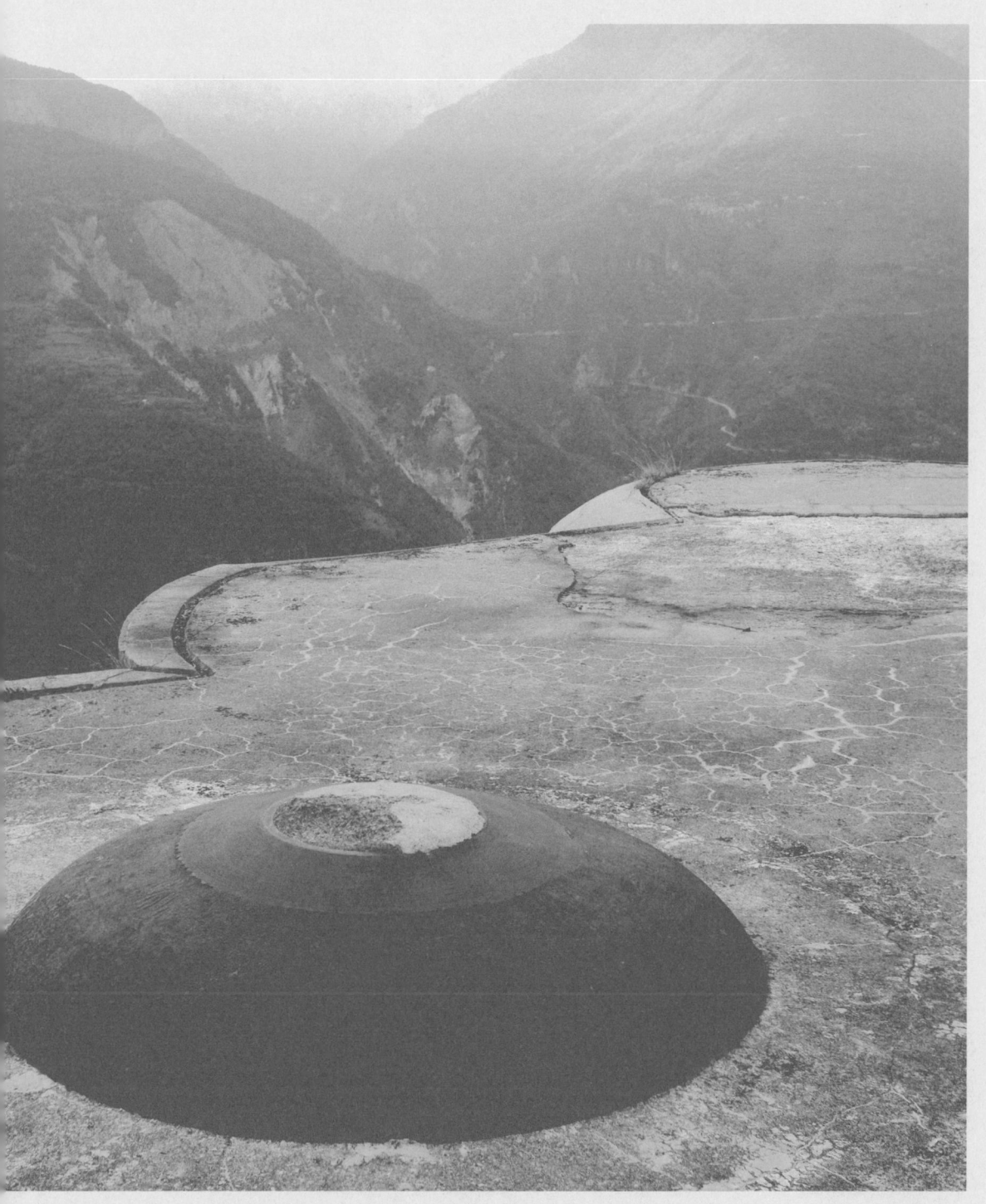

the most beautiful areas of France, like stray facts from a future passed, they are guilty reminders of things that we would probably rather forget; latent tendencies to attack and to destroy, old feelings of enmity and hate. But mainly, with these defences in the Alps, the feeling is of a great misplaced energy, of waste. The onslaught they were built to defend against never came.

Perhaps the psychological deterrent of these sentinels was crucial in barring this route of attack. But in hindsight it seems obvious that Mussolini sending troops over the Alps would never cause France's downfall. From our vantage point today, these fortifications might be seen as, if not a systematic folly, then a white elephant, and a testament to a failure of sorts.

It was not a failure of materiel: the fortifications of the Maginot Line (of which these are a little-known part) were never breached from Montmédy, near Belgium in the north, to the Mediterranean Sea. But war is a test of opposing imaginations as much as technologies; while the technology was more than adequate, perhaps in this case the French imagination proved too romantic, or simply unable to apprehend the real threat.

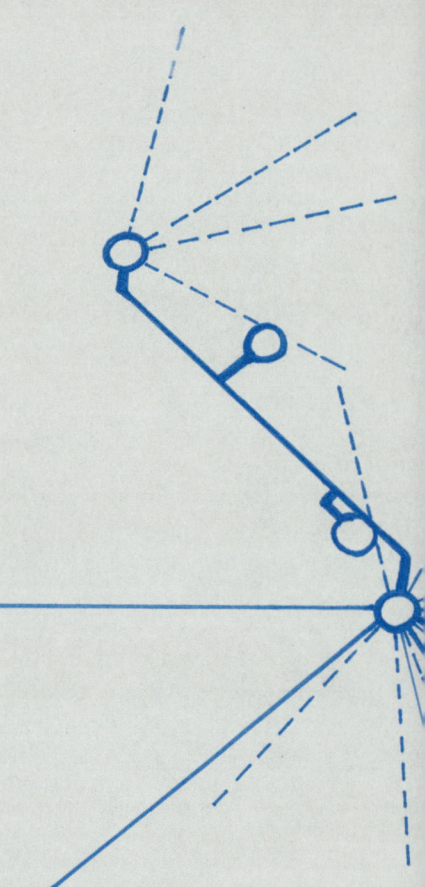

▲

We explored by bicycle first, over many separate trips riding further and higher until one Easter we found ourselves on unpaved tracks, carrying bicycles over snowdrifts towards forts guarding a border that no longer was. We ranged wider even as our targets became more focused, narrowed down to precise GPS coordinates. And with every drive and hike, every scramble through bushes or up snowless black runs during the worst ski season this millennium, the secret history of this deserted backcountry became a little clearer.

The Alps are shared between eight countries. They are at once a natural barrier between languages and nations, a contested liminal space in which boundaries are constantly slipping and a region unto themselves, where neighbouring peoples divided by a borderline often have more in common with each other than with their distant capital cities. The mountainous frontier between France and Italy is now the preserve of hikers, tourists and cyclists, but it has suffered repeated incursions, invasions and attacks since the dawn of history, when Hannibal led his army and his elephants over a high pass towards Rome. In valleys and on ridges from north to south are fortified towns, garrisons, lookout points and castles

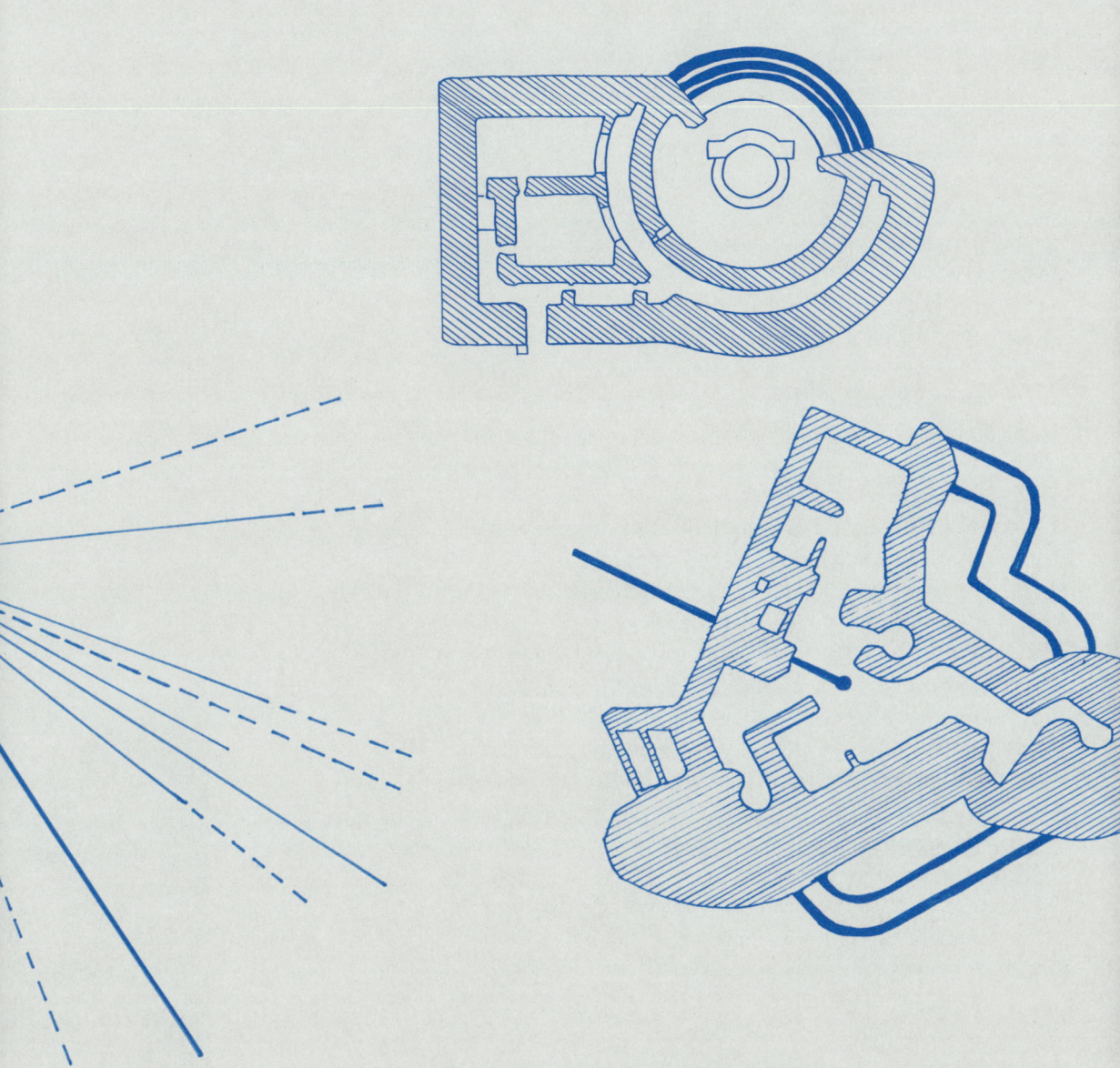

« Almost all had survived the war and seemed to be waiting for the next one, left behind by a race of warrior scientists obsessed with geometry and death. »
'A Handful of Dust'
JG Ballard, 2006

dating from the Middle Ages onwards, sometimes several kilometres from the current border. Hannibal's precise route over the Alps is only now being satisfactorily traced, but it was certainly north of the fortifications of the Alpes-Maritimes pictured here.

These mark a border that came into existence much later: it was not until 1860, during the birth traumas of the Italian nation, that the ruling House of Savoy gave Nice and its surrounding county to Napoleon III of France. Nice at that time was an Italian port to rival Genova, 200 kilometres to the east. After the cession, it was a beautiful French jewel sitting on azure waters near an undefended border, vulnerable to being taken back at any time. Not only Nice was threatened. Napoleon III had granted Vittorio Emanuele II continued possession of his favourite hunting grounds, which left the territorial agreement intrinsically flawed: the border, which might logically have followed the watershed along the peaks and ridges to the sea near Menton, cut back and forth across a sizeable valley, leaving several French towns isolated and exposed.

Barely a decade later, in 1871, France had been humiliated by the Prussians and had lost the Alsace-Lorraine region, and tensions were again rising with Italy. All these new borders with France's eastern neighbours needed defending. Starting in 1874, a system of forts was conceived and built by General Séré de Rivières of the *Comité de Défense*. These forts were the first generation of defences to use reinforced concrete – better able to withstand new types of shells and artillery. The northern parts of Séré de Rivières' creation saw action in the First World War, but the system in the southern Alps was never tested. Before that war, Italy had been allied with the German and Austro-Hungarian empires, but it eventually sided with Britain and France, and those forts on the Italian border, which had seemed so crucial, were sidelined.

˄

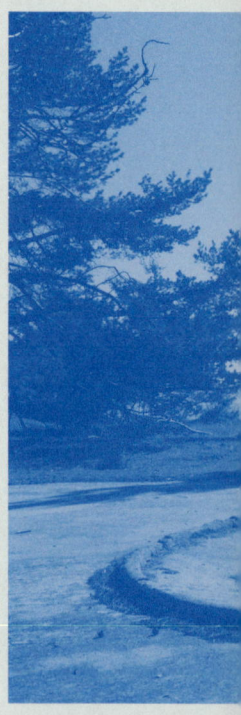

Even in a wild landscape such as the Alpes-Maritimes, it is rare to be somewhere that has not been shaped by human activity, or at least the human mind. Agriculture and hunting, settlement and travel, conquest and defence, enclosure and openness, ideas of the picturesque and the sublime. In time, the underlying artifice makes itself felt and you begin to question everything you see.

The French call these bunkers *ouvrages* (works) or *forts*. But at its simplest a bunker can be defined as a military structure for

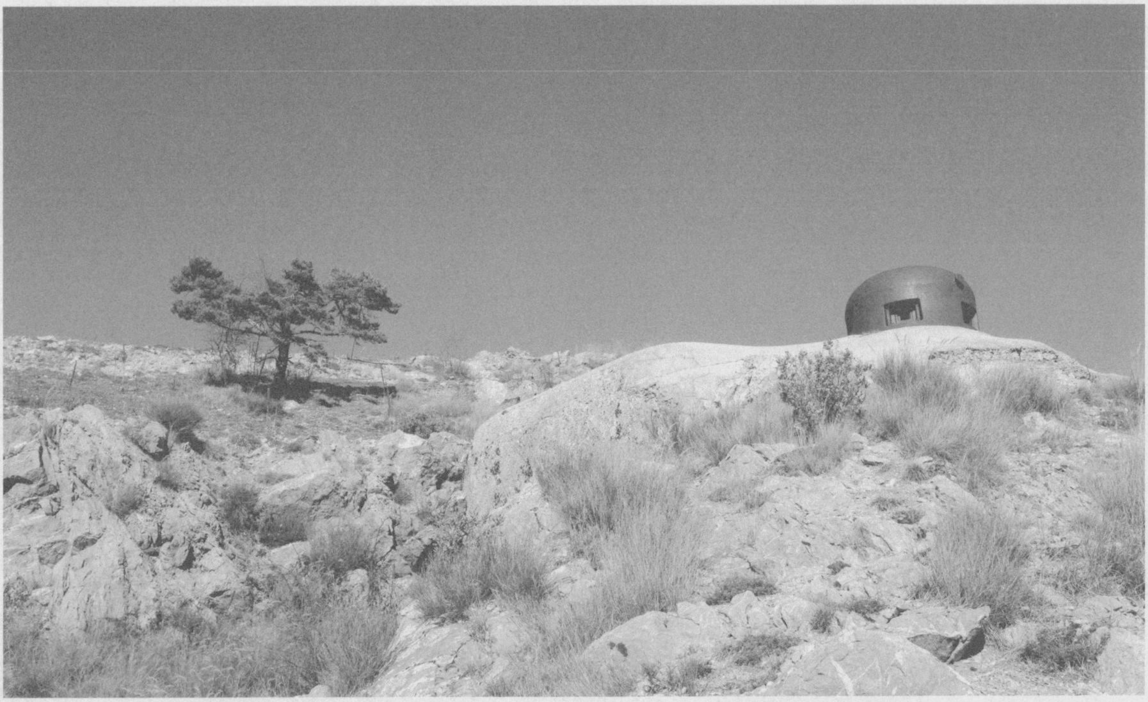

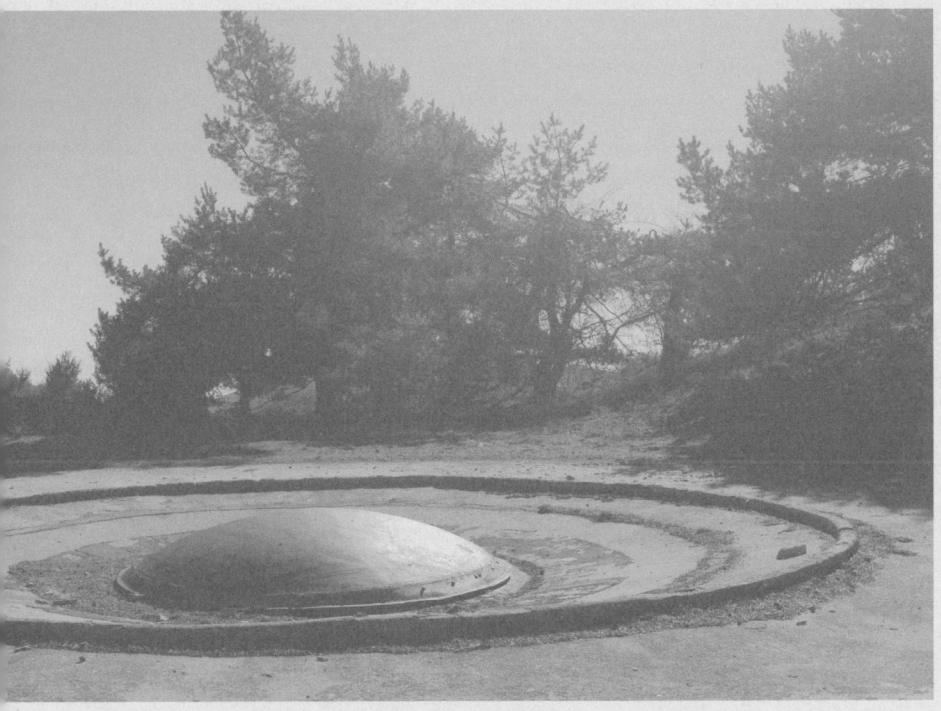

defence or protection that exists mainly underground – and that is what they are. At first, when you look for them, there isn't much to see. Their soft, rounded edges were designed to avoid throwing unnatural shadows across the landscape, and the concrete was mixed on site using local aggregate, so the *ouvrages* are all subtly different in colour, blending in with the rocks that surround them. Sometimes the walls bear the marks of the shuttering into which the concrete was poured and, like a Rachel Whiteread sculpture, record or even mimic the grain and the knots of the now absent wooden planks that formed them.

Now, more than 70 years after their construction, they have softened into the landscape even further. Spend long enough looking, however, and your vision changes, becomes accustomed to the bunkers' hidden logic. An entrance on the western side of the hill naturally and inevitably leads to casemates or *cloches* – domed cast-iron fixtures for shooting or observation – on the eastern side facing Italy. An emplacement on one side of the valley means there will be another on the opposite, with empty embrasures staring both in the direction of the threat and across to its sister structure (they were all designed to withstand direct hits from other Maginot guns, allowing each to provide covering fire in the event of an attack with no worries about collateral damage).

Gradually, you sense the vast empty networks of tunnels below your feet, and begin to imagine these deserted places busy with men, humming with life.

Mostly, though, as you draw near, you experience a gathering apprehension. Once you realise the sheer number of installations crowding the hills, it occurs to you that the road or track you are on, which at first seemed innocent, was originally built by the military, that it too has played a part in this drama. And, whether you are driving, riding a bicycle or hiking along a marked path, you realise that you are under surveillance. The dead eyes above are tracking your progress, ready to fire down upon you. You have been drawn into, implicated in, the conspiracy.

▲

During its construction, the Maginot Line was thought to be a triumph of French defensive engineering. The first and most famous part of the Line comprised concrete fortifications and weapon installations extending the length of the border with Switzerland, Luxembourg and

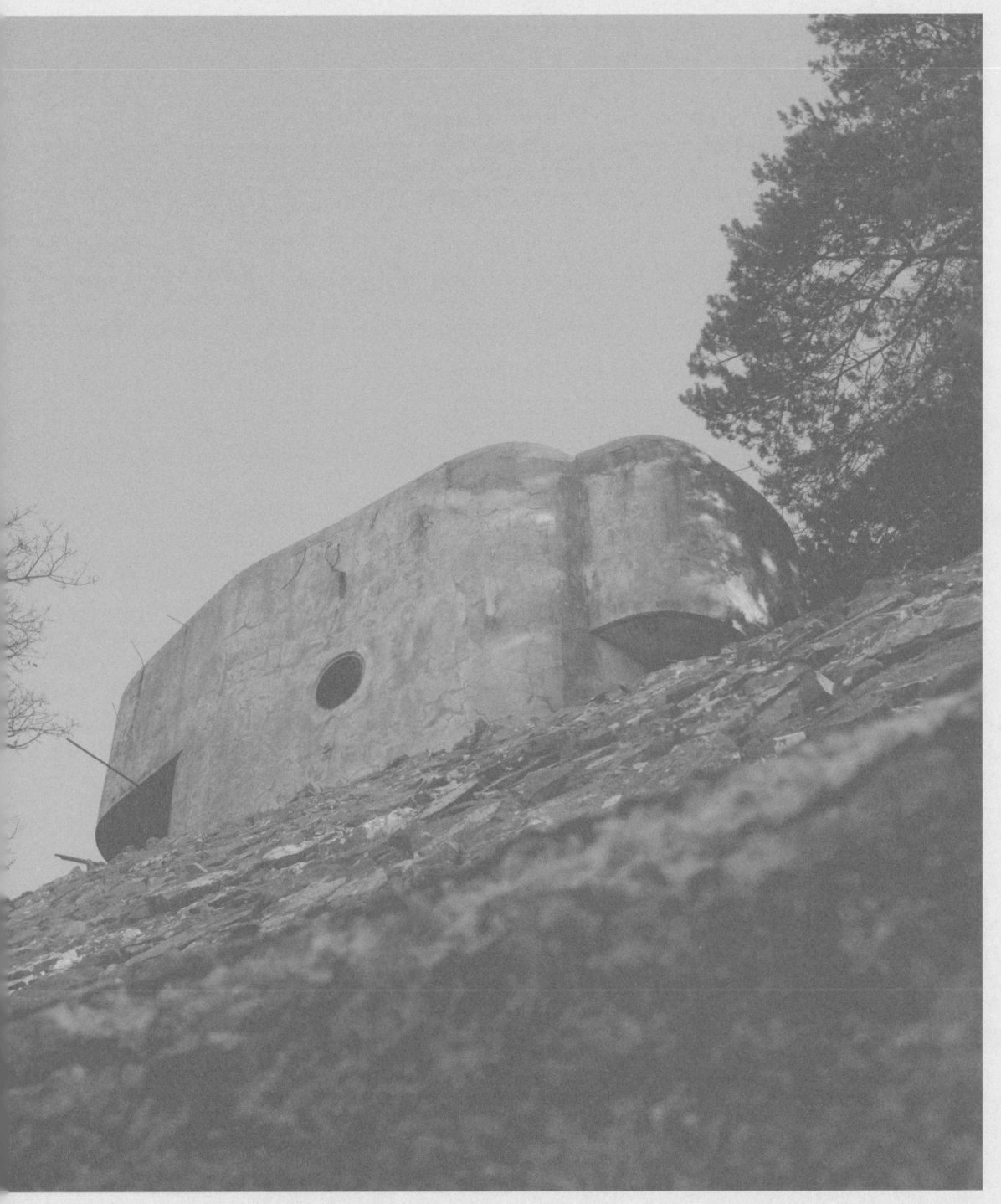

Germany. It was the brainchild of politicians Paul Painlevé and André Maginot, and its founding vision was of the prospect of another war in which static, defensive tactics would prevail.

The Alpine Extension, also known as the Little Maginot Line, is less well known, and, given the French conviction that the main future battlegrounds would be in the north-east, was not initially a priority. Although Italy had been an ally in the First World War, with Mussolini ascendant in the 1920s it became clear that the Italians were not content with their territorial settlements at Versailles. Increasingly irredentist, he threatened to reclaim the Savoie region, Corsica and Nice. The work schedule in the south was accelerated: the great fort at Rimplas (p.6 & 80) north of Nice, was the first site to break ground anywhere on the Maginot Line, and was among the first to be completed.

Despite budget shortages hampering construction, the *Secteur Fortifié des Alpes-Maritimes* (SFAM) became the most heavily guarded sector of all the Alps. It was also the most varied, reaching almost to the Col de la Bonette, even now the highest paved road in Europe. Its architects and engineers had to be more inventive than in the flatlands of the north, where construction was more standardised, and they produced some extravagant forms in response to the challenges of the terrain.

Many of these high-altitude *ouvrages* could only be built during the brief Alpine summer, but it takes time to haul multiple machine gun *cloches*, each weighing 15 tons, 25 kilometres up a dirt track to a building site. Consequently, numerous outposts were unfinished at the outbreak of war in 1940. Many of the original plans were scaled back, modified and made less expensive but the extent of what was achieved in the mountains remains impressive.

However, the Line also extended south to protect the densely populated Côte d'Azur. That meant our bunker research was not confined to the mountains. The basin around Sospel, one hump away from the sea, was especially heavily guarded. Here, gun turrets encircle the town, bunkers hide in crannies of rock and *cloches* stand guard in olive groves and in between the tomato plants in allotments. Closer to the coast, the remains have been assimilated into the sprawl. They camouflage with the rock in a restaurant car park in Sainte Agnès (p.34), nestle between the expensive villas of Cap Martin (p.12), pop up in a suburban park in Roquebrune (p.8 & 14). You might pass them every day and not give them a second glance. They are still watchful, entirely visible and yet invisible, neutralised and domesticated. The *unheimlich* has been made homely.

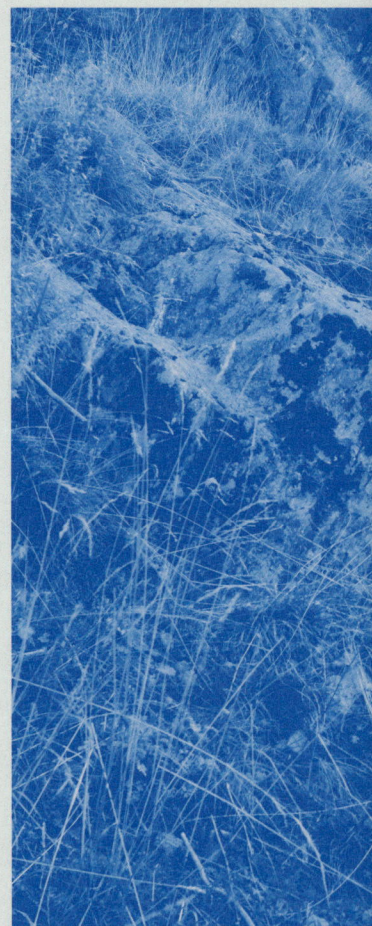

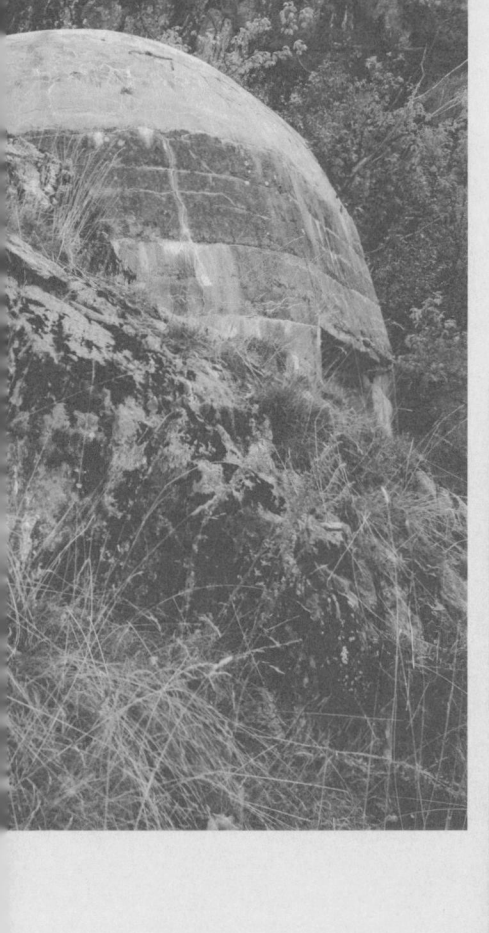

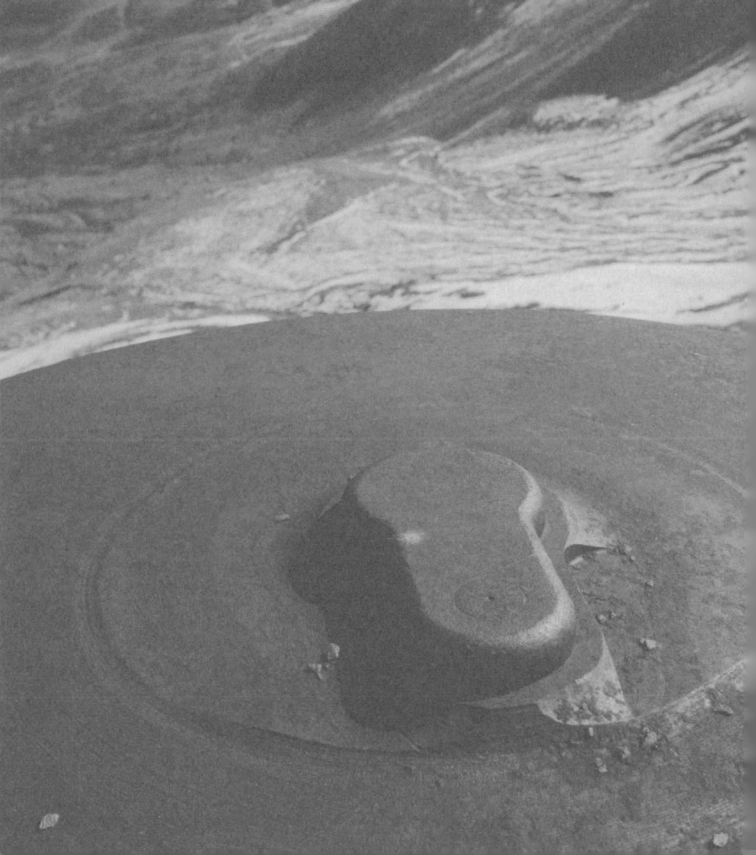

When you cycle a mountain road it winds like a ribbon, wrapping itself across the contours as it climbs. Bunker research meant approaching the terrain more laterally: finding the sightlines across, the hardest points to reach – understanding the Alpes-Maritimes as a military conundrum as well as a sporting playground. The landscape becomes a game of axes and channels, weak points and redoubts, penetrations and surprises.

Three main river valleys needed defending to secure Nice from the Italian threat. One, the Tinée, leads all the way north to the remote Col de la Bonette, whose fortifications were the highest of the whole Maginot Line (although on the border of the Alpes-Maritimes, these actually belonged to the SFD, the Dauphiné fortified sector). The second, the Vésubie, is further to the south and east, and in its middle reaches is wide and relatively open.

The constellations of buildings in these valleys, and across much of the rest of the terrain, was broadly similar. First, there were small *avant-postes*, usually as close to the border as possible. Manned by a handful of soldiers, these lookout posts could be several hours' march up goat tracks, and consisted only of a fortified position, a shelter and some trenches. In the more remote areas, the *petits ouvrages* were made up of mixed infantry and artillery blocks overlooking key passes and border areas, or occupying vantage points defending valley roads. They housed maybe a hundred men and were equipped with grenade launchers, mortars and machine guns. Then, in the most strategic sites, or where the threat was greatest, the *gros ouvrages* would contain large-scale weapons or barracks for hundreds of men, lift shafts connecting multiple underground levels and small railways for moving ammunition and supplies. The *ouvrages* had sophisticated air filtration systems in case of attack by poison gas and were often positioned over springs, to assure their water supply. With generators, bunkrooms, latrines, storerooms and kitchens, each was designed and stocked to be self-sufficient for three months of underground living.

Further down the Vésubie and the Tinée, towards their confluence with the Var (a river valley that was effectively a highway to the coast) were two 19th-century *chiuses*: barracks and gun emplacements hollowed out of the vertical walls of the gorges at their narrowest points, with retractable road bridges to cut access. These troglodytic dwellings were a legacy of General Séré

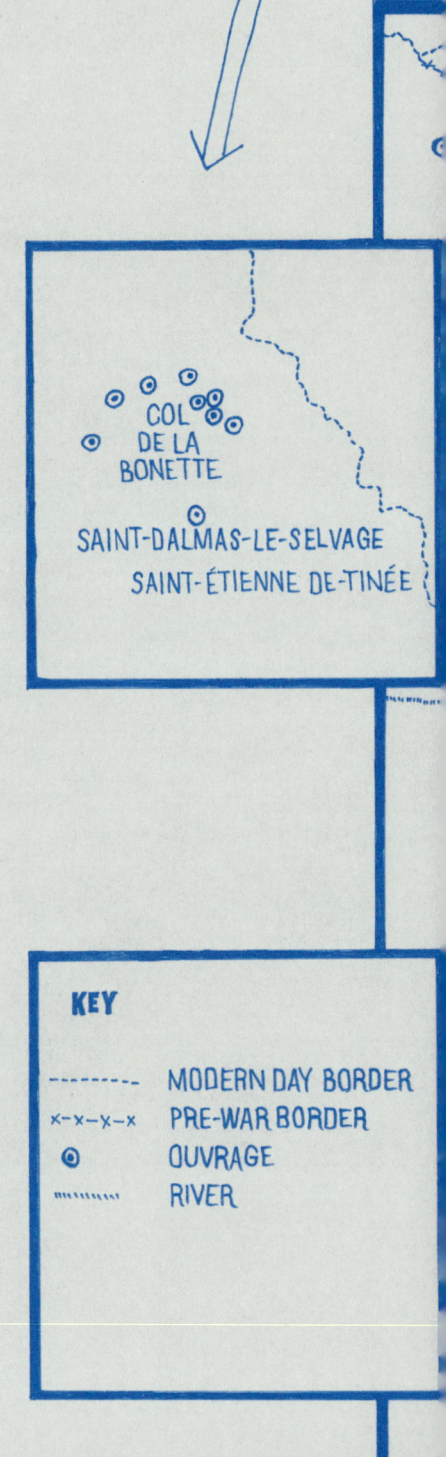

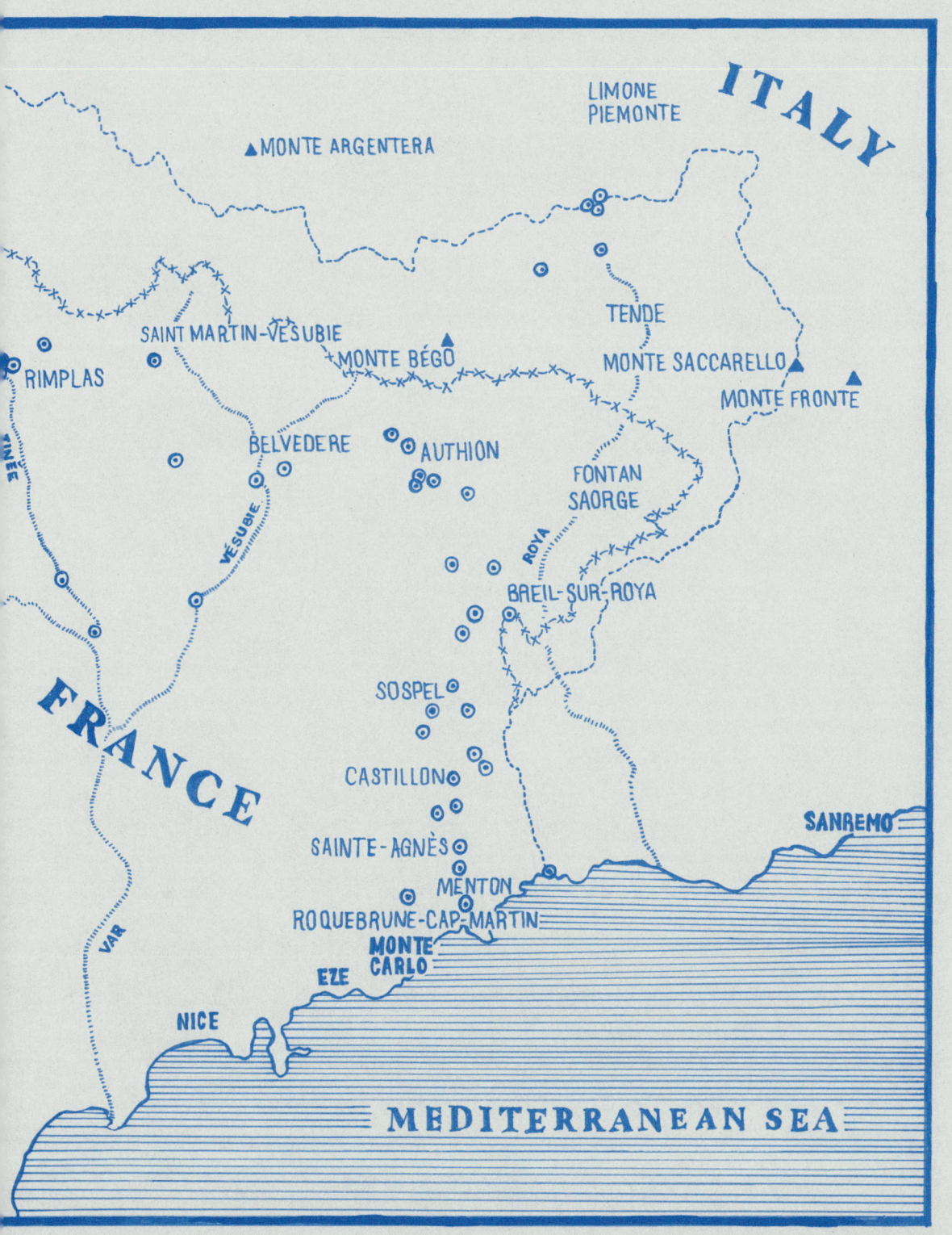

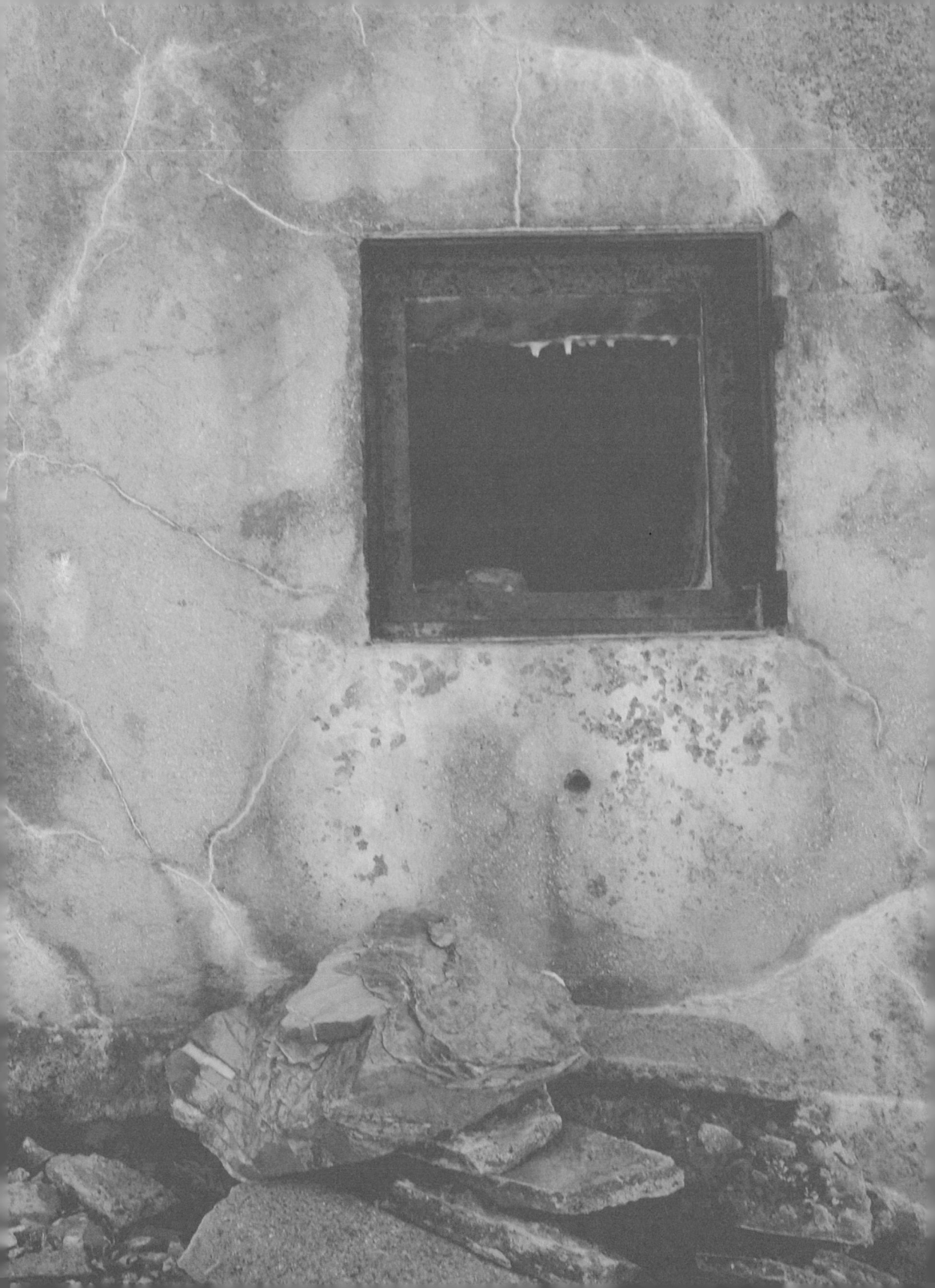

de Rivières 50 years before the Second World War, but they were repurposed and used in the 1930s.

Many of the older fortifications had a second life: if the 19th-century forts were well positioned, they were reoccupied; some were also modified with 20th-century concrete appendages to the original structure.

▲

Furthest south and east, the third valley, the Roya, was a complicated case. Thanks to Napoleon and Vittorio it had always been part French, part Italian, with French towns, notably Breil-sur-Roya, stuck out nude and defenceless between two parts of Italy, connected by a single mountain road to the rest of France. At the top of the valley above the town now known as Tende was the Colle di Tenda, for years a key pass on the salt route from the coast up to the Savoy seat of Turin.

The bunkers along most of the Roya were therefore Italian, part of a system known as the *Vallo Alpino* (Alpine Wall) – a defensive mirror of their French counterparts a few kilometres away. Many of these are inside the present-day French border: stranded in enemy territory, their past is a lost history. The Italian bunkers are appreciably different. Their forms are more rounded, embrasures more menacing, their construction at once more sophisticated and more primal. Each site has a particular atmosphere, and at the Italian bunkers it is always heavier, darker and more oppressive. Something in the air testifies to bad things barely buried by time passing – secret crimes – and it feels as if the birds do not sing.

The Italians were also more creative with their disguise, enveloping fortifications in rough stone textures. Some were even made to resemble a kind of artificial rock face, using chicken wire, ceramics, sprayed-on cement and paint (pp.76-7), creating, it seems, a set for a war film that never went into production, or a prop for the re-enactment of a battle that didn't even happen.

Today, the whole of the Roya valley has the air of a no-man's-land. It is neither France nor Italy, but rather just what it is. It probably always was. Here, the language spoken was a dialect known as Tendasque and the peasants tended market gardens on the terraces above the dead-end villages, or took their sheep, goats and cows high up into the mountain pastures where they lived for the summer months surrounded by the remains of a bull-worshipping Bronze Age cult.

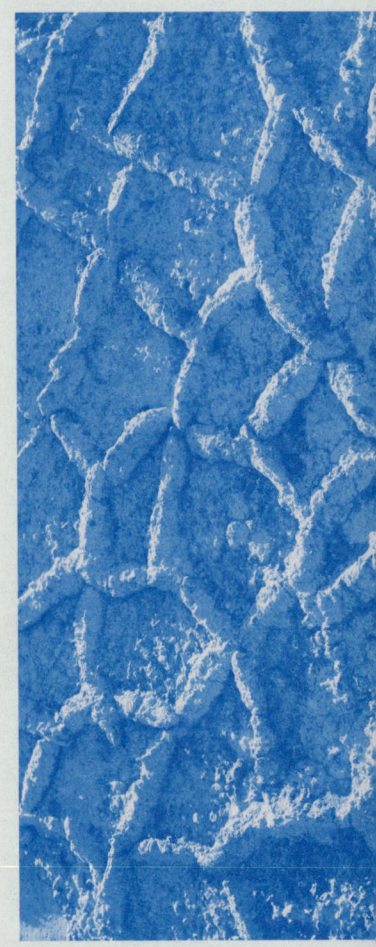

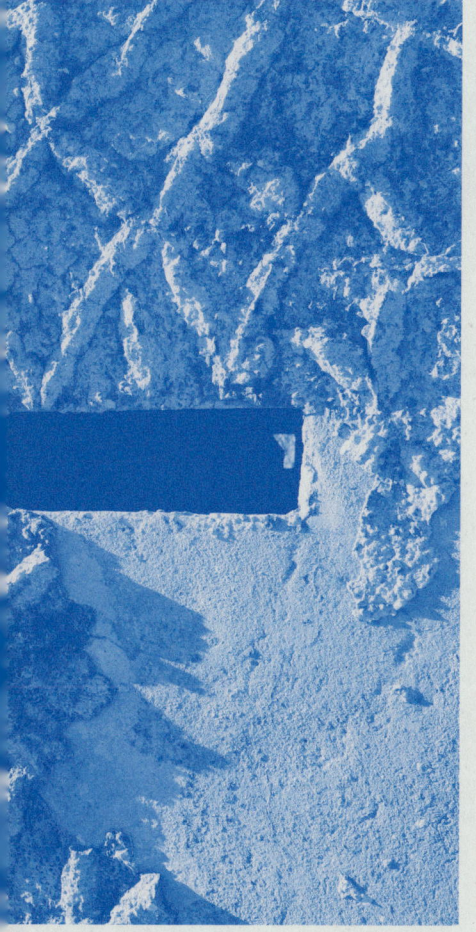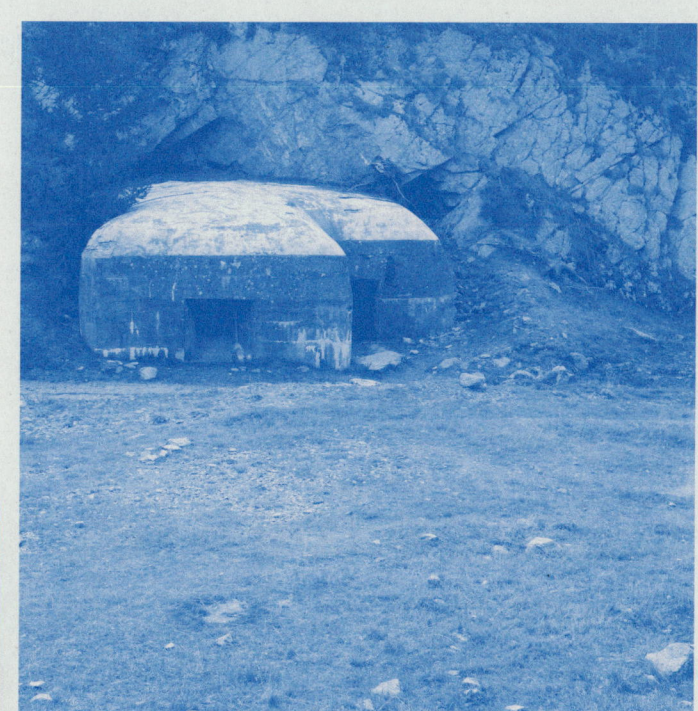

It's just one example of the cockeyed logic of the war in these parts – that a people neither truly French or Italian was forced to pick sides, flee their homes when the Italians invaded, endure occupation by the Germans or deportation to Turin after the Italian armistice, and then finally to become French again (or for the first time) in 1945.

Another example: some of the contracts to build French fortifications in the Alpes-Maritimes were won by Italian firms. Not only did the enemy know where the defences were, they had seen the blueprints and poured the concrete onto which the guns were mounted.

▲

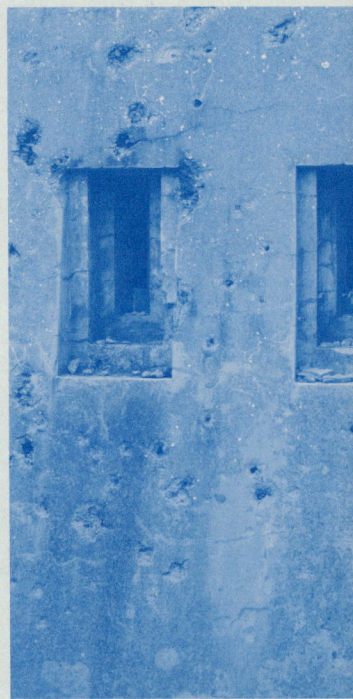

Not that this, in the end, mattered much. The Battle of the Alps was short and decisive, since the main action happened elsewhere: Hitler's assault began in May 1940 and France was overrun in the Ardennes, far to the north, where no Maginot defences had been built. It wasn't until 10 June that Italy declared war on the Allies, and even after that the Alpine front was quiet (although the border post at Menton's Pont Saint Louis (p.4) received fire a quarter of an hour before the war's official start at midnight, and was still being shelled several hours after hostilities ended on 25 June).

Only when France had signed its armistice with Germany did the Italians really strike. The main attack in the Alpes-Maritimes saw Isola, Breil, Fontan and other mountain towns occupied; on the coast, the border post was overwhelmed and the Cap Martin *ouvrage* received heavy fire as Italian troops took the town of Menton. But many forts saw little or no fighting. The SFAM motto – *On ne passe pas* (None shall pass) – held true.

On the stroke of midnight, as the French surrendered, lightning hit the radio antenna on one of the *ouvrages*, and the men within thought the Fascists were getting in one last shot before the ceasefire.

After the armistice, the Italians took control of the bunkers but, since they served no strategic purpose, did not use them. When the Italians left the Axis in 1943, the Germans occupied the Alpes-Maritimes, and did not use the majority of them either. Only a few, such as Roquebrune and Agaisen (p.16 & 106–7), were manned by the Nazis, and these were vandalised or blown up as they left. At many of those that had remained empty, the keys were simply handed back.

Most of the sites had been more than 10 years in the conception, planning and construction, 18 months in active service and 15 days,

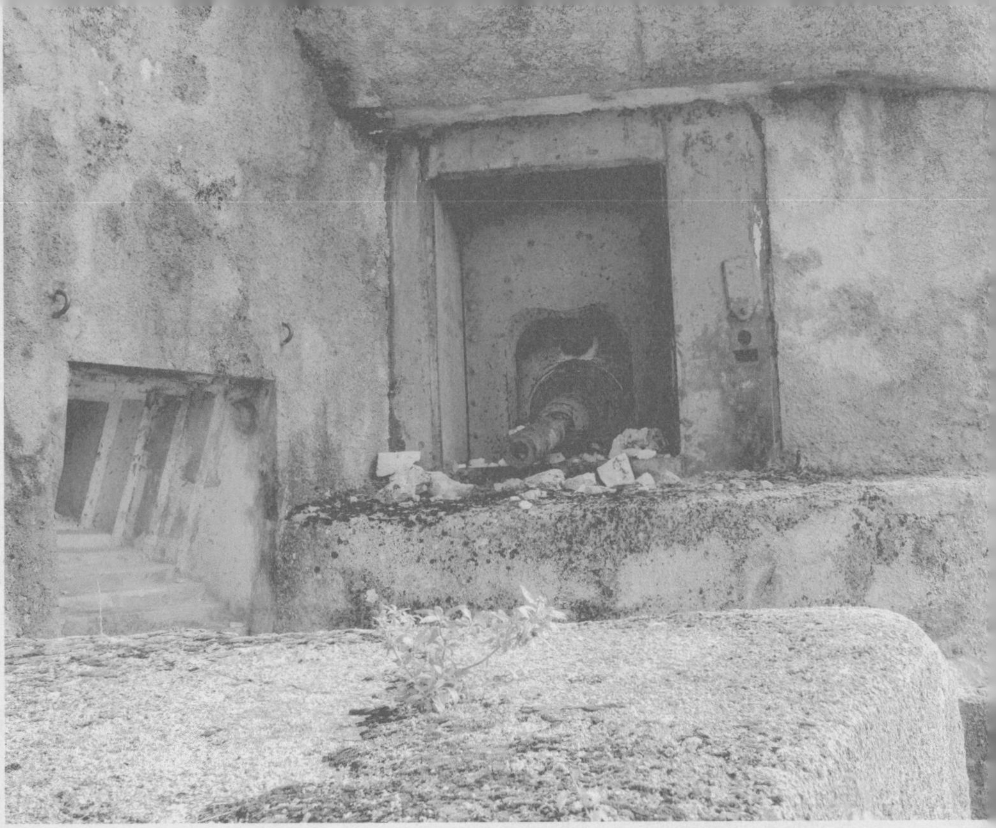

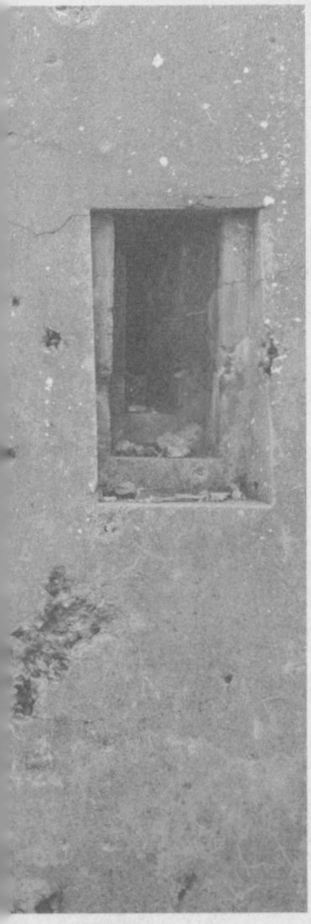

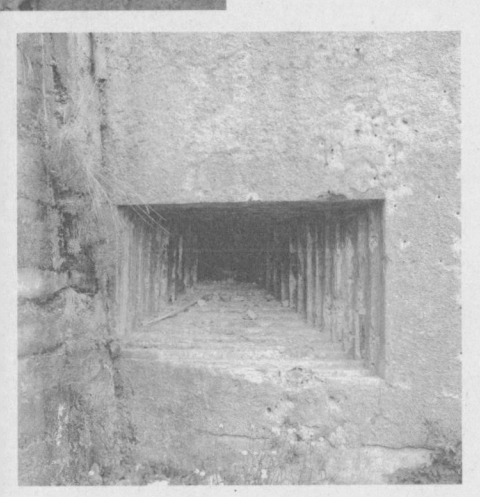
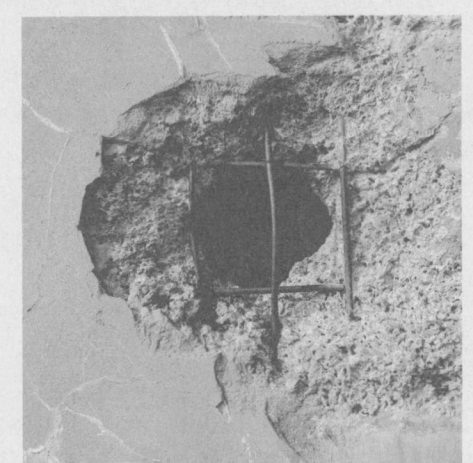

at most, in battle. Hundreds of millions of francs, thousands of tons of concrete, hundreds of men stranded in dark holes high in the mountains through desolate winters, for nothing.

▲

One of the ironies of the war in the Alps was that the only Maginot fortifications to see heavy action were those attacked by the French themselves.

While Paris was liberated in August 1944, it took until spring 1945 for the Free French and the Allies to make a final push to the Italian border. There, they encountered a problem: the Authion massif. The Authion is an almost perfect fortress of 2,000 metre peaks overlooking key border territories. It had once before been a battleground, in France's revolutionary wars, and it was where the Germans dug in for their last stand on French soil. The retreating troops occupied both the Maginot bunkers and Séré de Rivières forts and, since these were designed to defend against Italy, not France, modified them to cope with this new threat of attack from behind.

Only a concerted effort by Charles de Gaulle's Free French in April 1945, supported by American tanks and British commanders, prised them from their stronghold, and there was considerable loss of life taking back this last bit of occupied territory.

Tende wasn't liberated until 5 May 1945. However, given it had historically been an Italian town, 'liberation' is not the right word. De Gaulle in fact used this final assault on the Authion to make a play for the strategic Tende pass and, three days before victory in Europe, he achieved it. Tende became French for the first time in a hundred years only upon the signing of the Paris peace treaties in 1947.

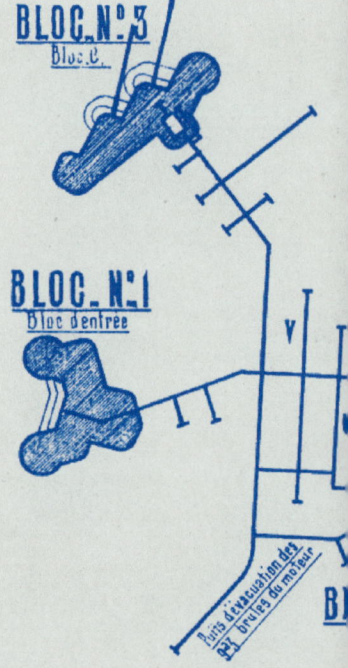

▲

Today, the Authion is quiet, although the ruins still upon it, like nowhere else in Western Europe perhaps, bear the scars of the heavy fighting they endured.

Elsewhere, the bunker sites have mellowed as they return to nature. The winches controlling the airborne supply lines have seized, the camouflage paint has faded, the timeworn surfaces are stained with lichen and the *cloches* have turned a deep rust brown. Lime is leaching from the concrete, forming stalactites that overhang the inverted ziggurats of the embrasures. As the bunkers

« **The bunker is the last theatrical gesture in the endgame of Occidental military history.** »

Bunker Archaeology
Paul Virilio, trans. 1994

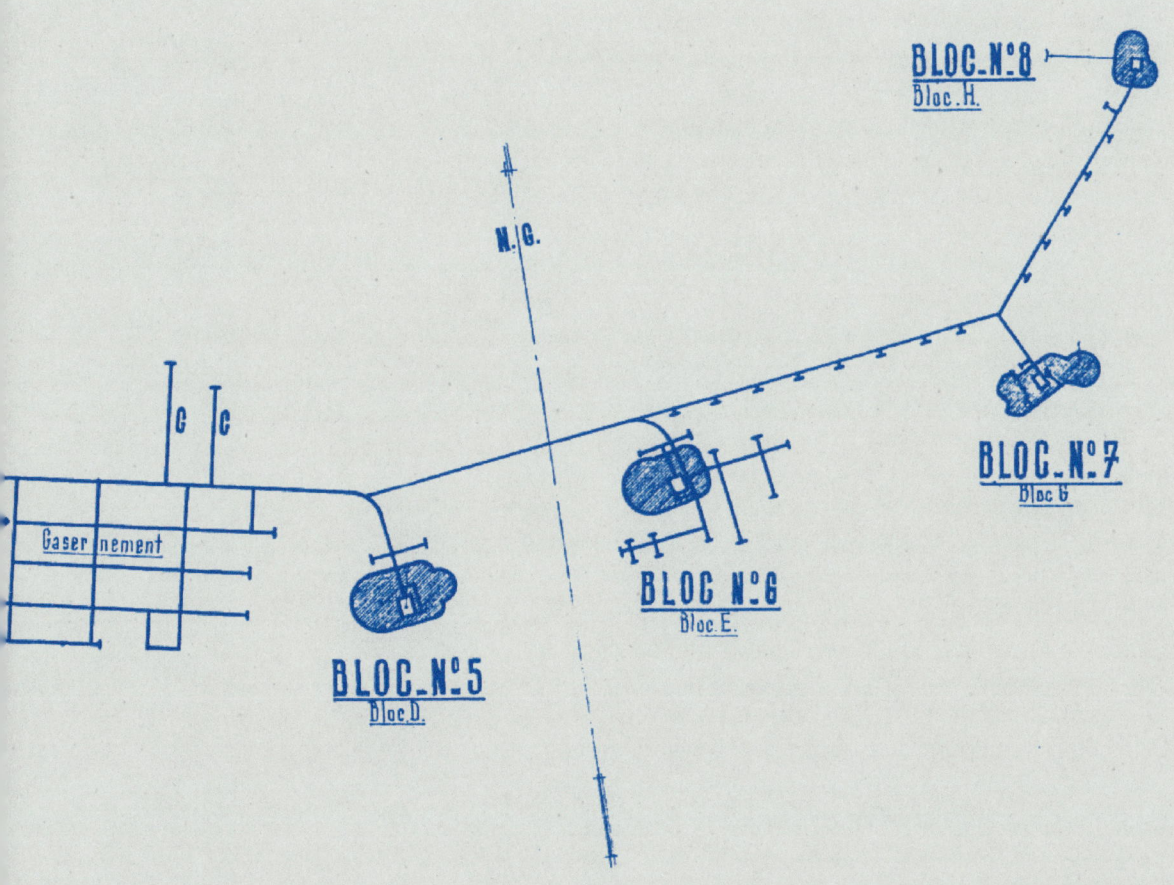

gently crumble they are becoming more difficult to see, and their squat shapes seem less dissonant with the soaring peaks, even complementing the rise and fall of the land around them. While some still dominate the landscape, others are now hidden in thickets or buried in brambles. Scrub oaks and pines have reclaimed the ground, blocking sight lines. There are still supply routes and vehicle tracks, but only walkers, cyclists or the odd chamois trespass here.

▲

Unlike the mountains, which speak of an eternity, the bunkers speak of a moment – the instant of a terrible and decisive impact – a moment, moreover, that never arrived. They may seem less and less out of place, but they will forever be out of time. We have moved away from the threat of an all-consuming war between countries or blocs; these fortifications won't save us from terrorists or cyber attacks. And although these borders are now patrolled again, after Schengen rendered them meaningless for so many years, these bunkers can do nothing to help solve Europe's refugee crisis.

Some of these relics of a bygone era have been rehabilitated, fitted out anew and opened in the summer months for guided tours, but most, and particularly the remote sites away from villages, have been left to a slow decay. What else can be done? How, precisely, do you remove a structure with concrete walls almost three metres thick that was literally designed to withstand massive explosions?

Visit them now, and it's easy to imagine the bare outlines, at least, of the misery of living within, under the oppressive threat of bombardment and death, and to feel the contrast between the dark discomfort and the glorious views without. André Maginot himself described them as a 'subterranean fleet', and below ground they resemble nothing if not the cramped quarters of a submarine. At the highest sites, in winter, entry was through a kind of periscope ladder that covered the main door and rose above the snow; in the 19th-century barracks close by, planks were laid from roof to roof, over the spaces between the buildings, so that men could walk around beneath the snow. After too long below ground, working in a day-and-night pattern of consecutive eight-hour shifts, men would complain of 'concretitis'. In some Maginot forts, mandatory group sessions lying naked under ultra-violet lamps took place, to prevent deficiencies caused by the lack of light.

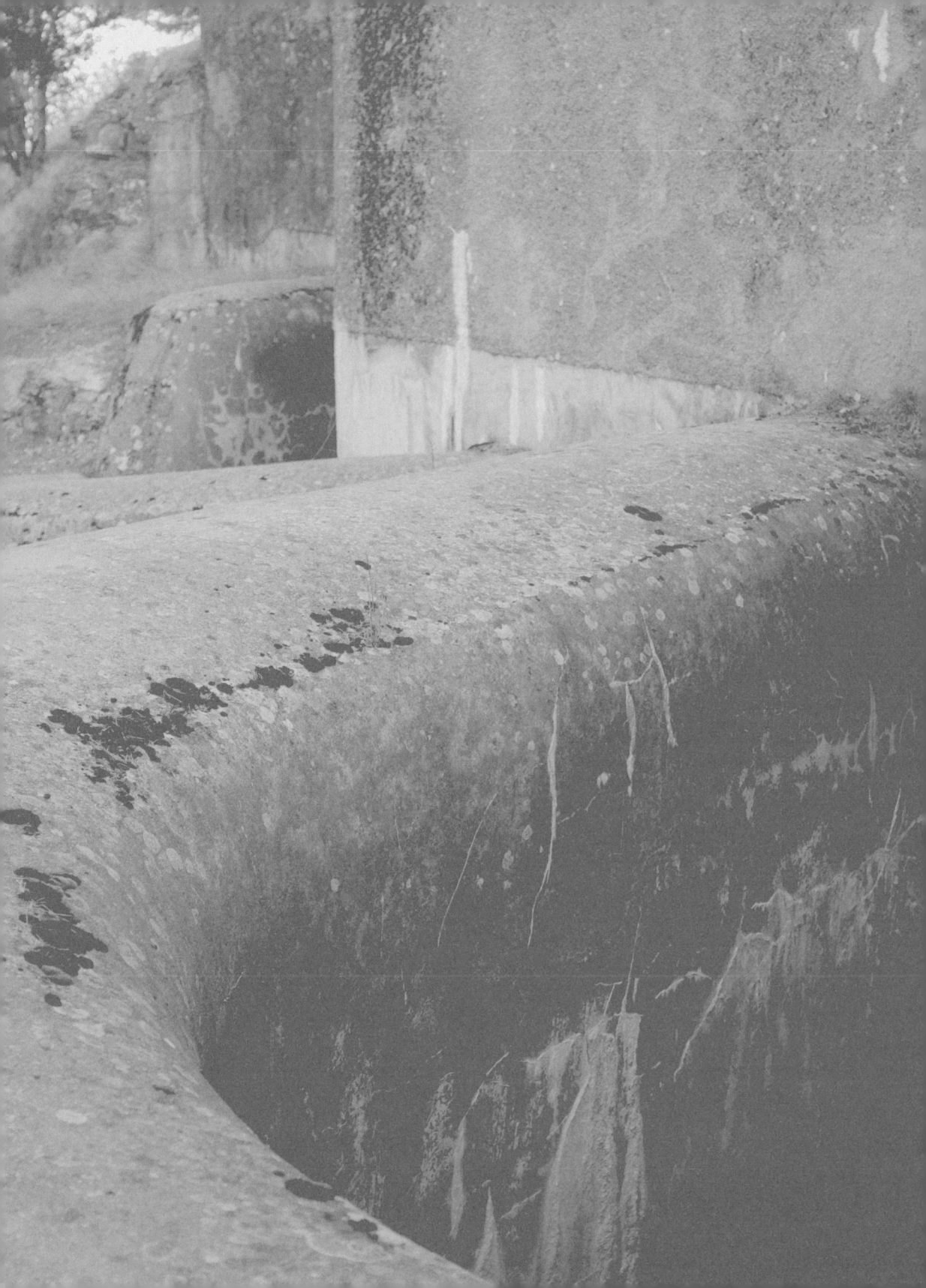

But life can't have been all bad. Many of the denizens of the bunkers were local men, and they would supplement their rations with fruit and vegetables from the fields. There were late-night card games and tuck shops stocked with local alcohol; at the *ouvrage* on the Col des Banquettes they kept two pigs – one called Adolf, one called Benito. A posting here away from the real horrors of war, surrounded by orchards or Alpine meadows, must have, in some ways, seemed lucky.

Seen now, in the soft spring sunlight of the Côte d'Azur, they are almost peaceful, at rest. Even 10 years ago, people writing the history of these forts could collect first-hand testimonies of life within. But there are now no longer any soldiers who manned them still alive, so this bunker research captures them poised between memory and oblivion. It is sad to think these war relics will long outlast the newer concrete leisure constructions – the hotels, swimming pools and villas – on the coast, but with every passing year, every freeze and thaw, they too inch ever closer to dust.

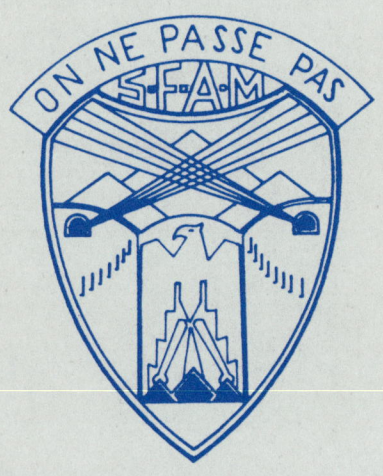

Following pages
Vallo Alpino
Italian positions
Isola

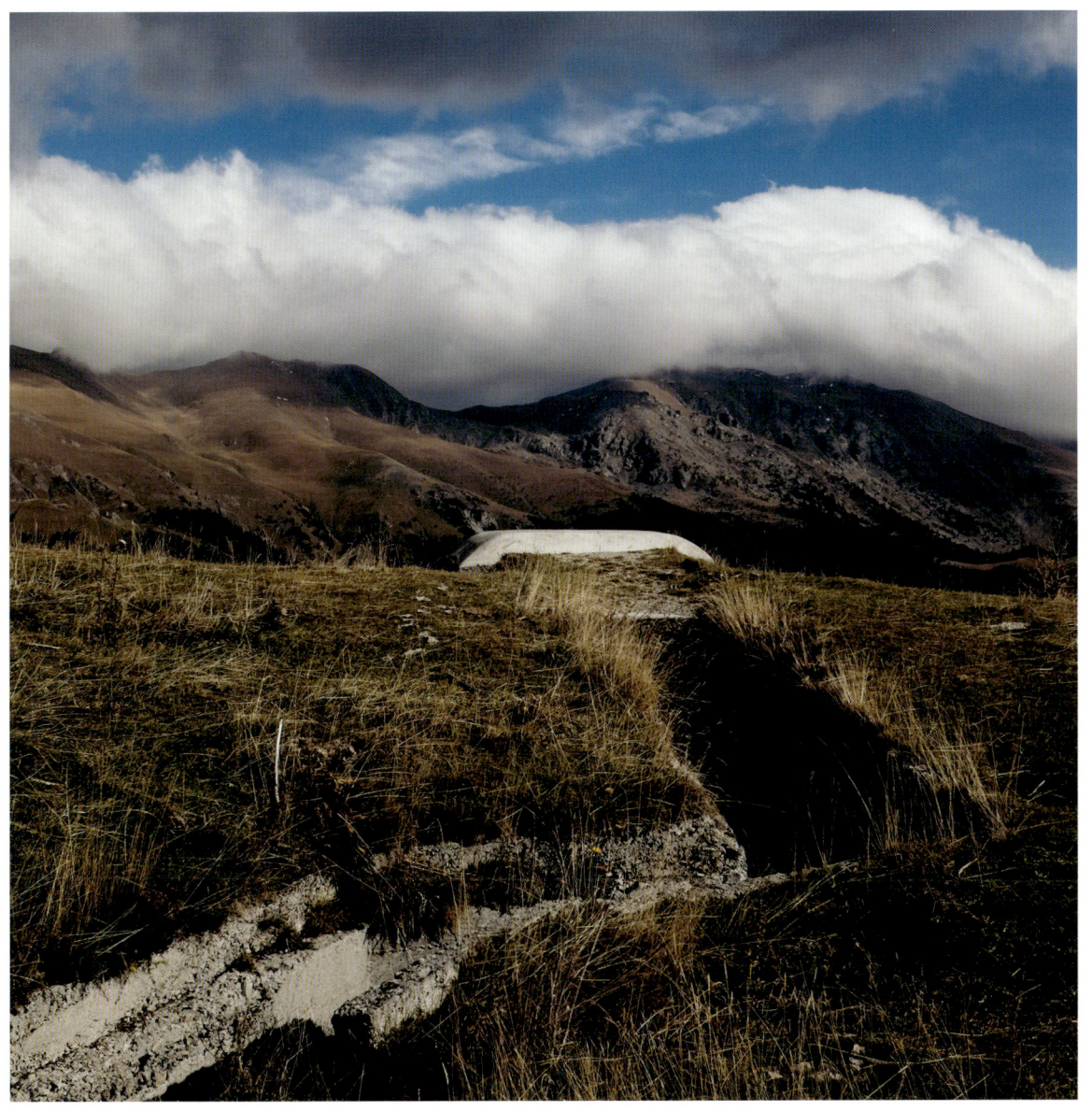

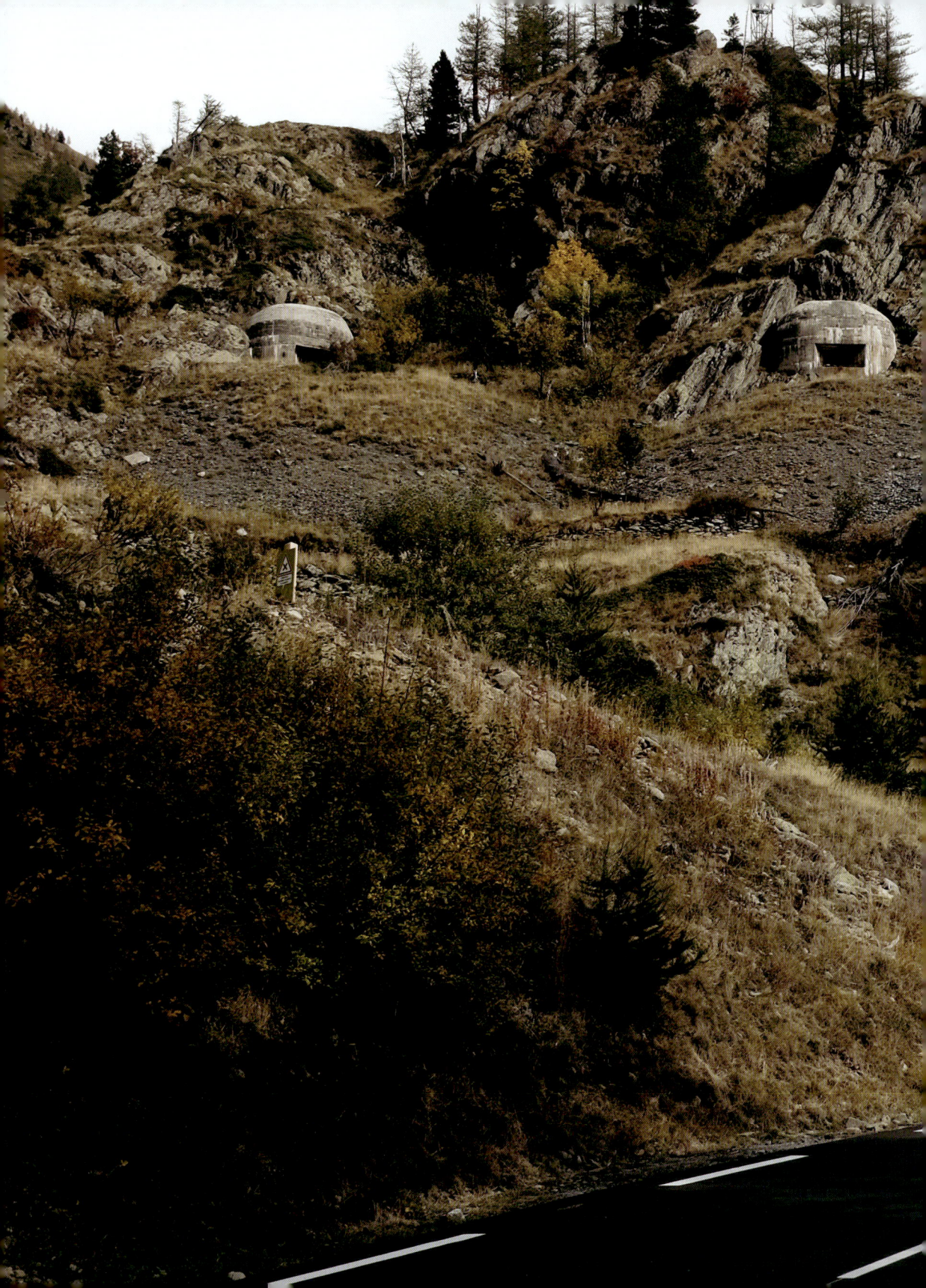

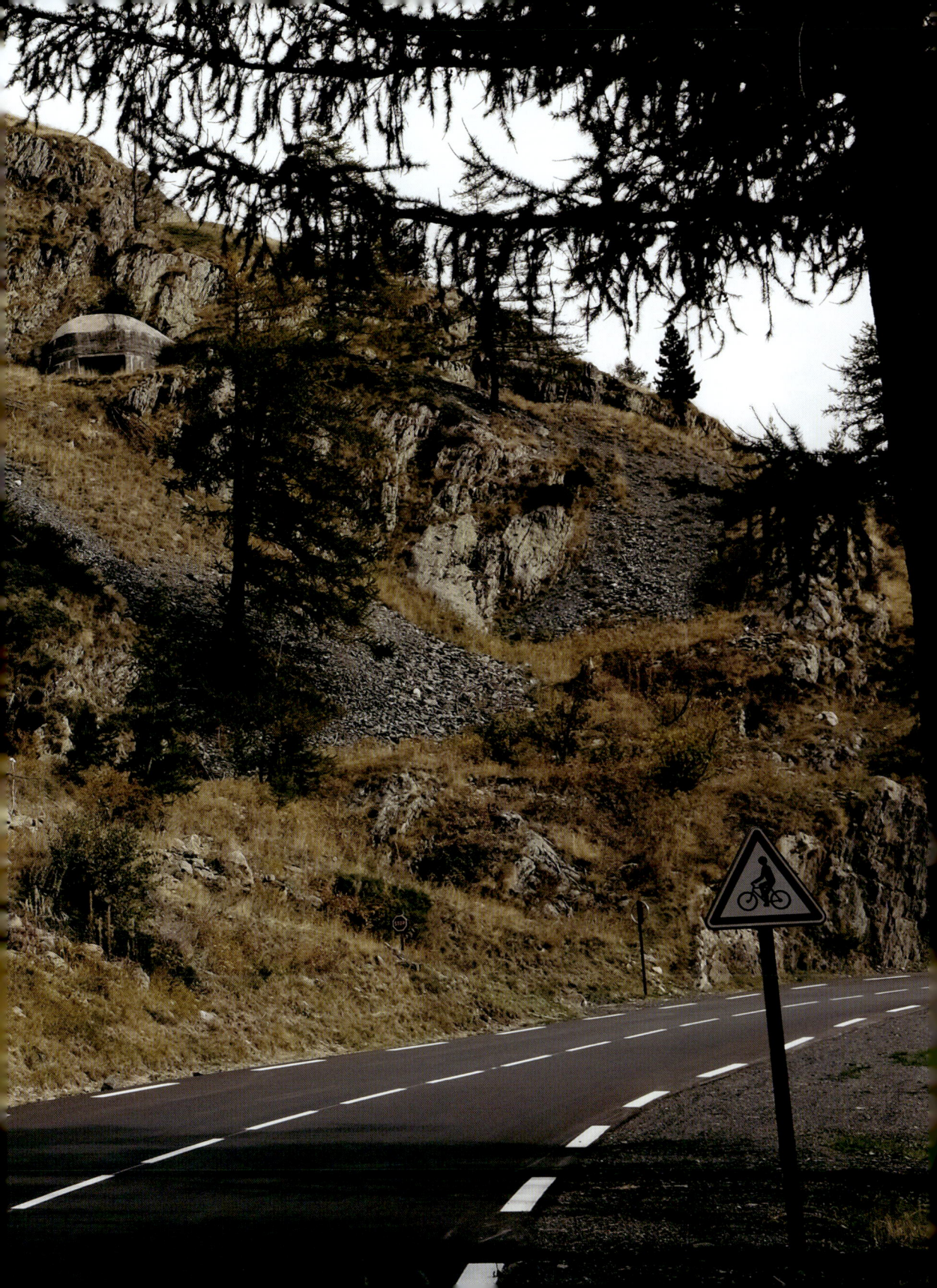

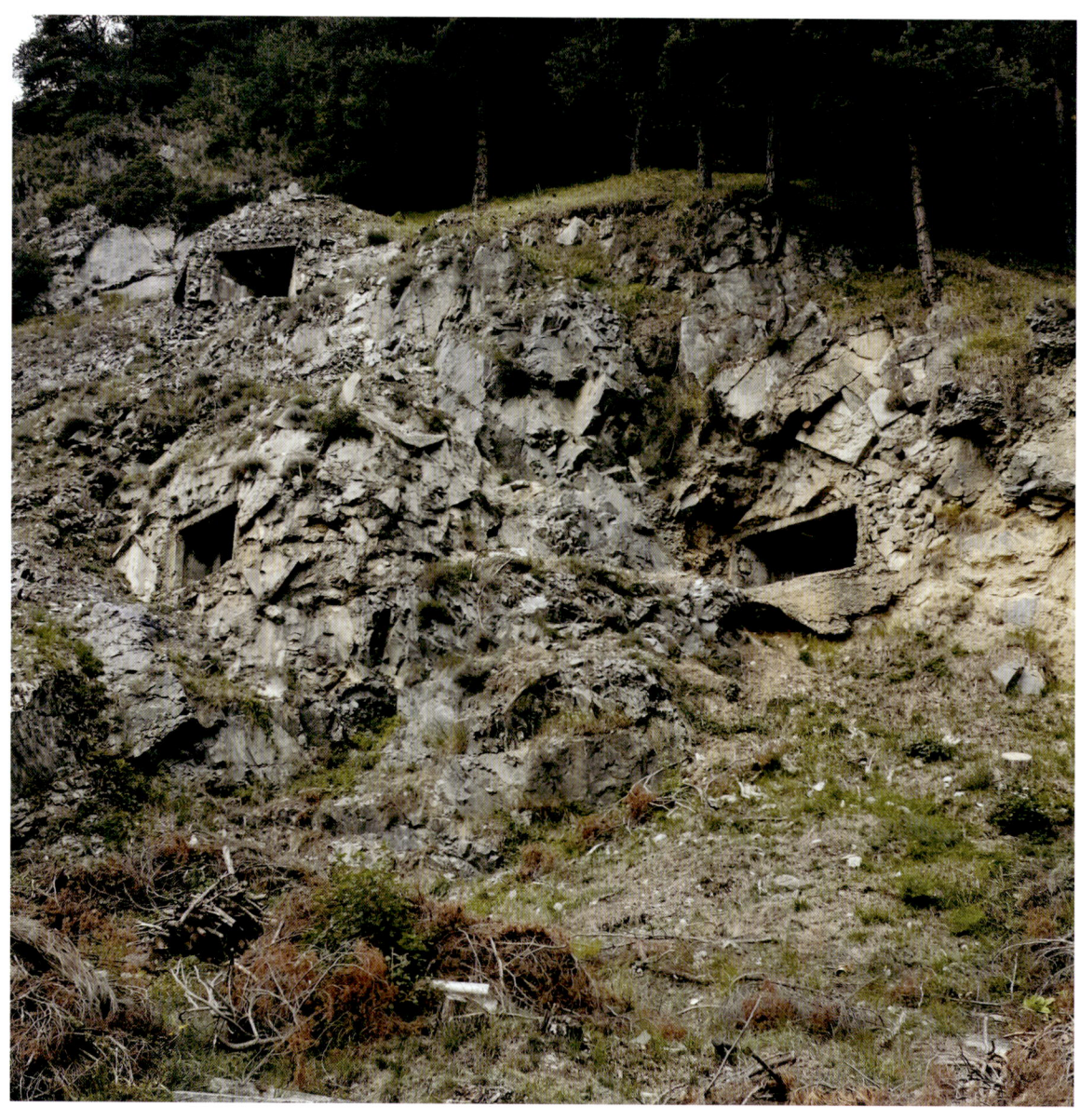

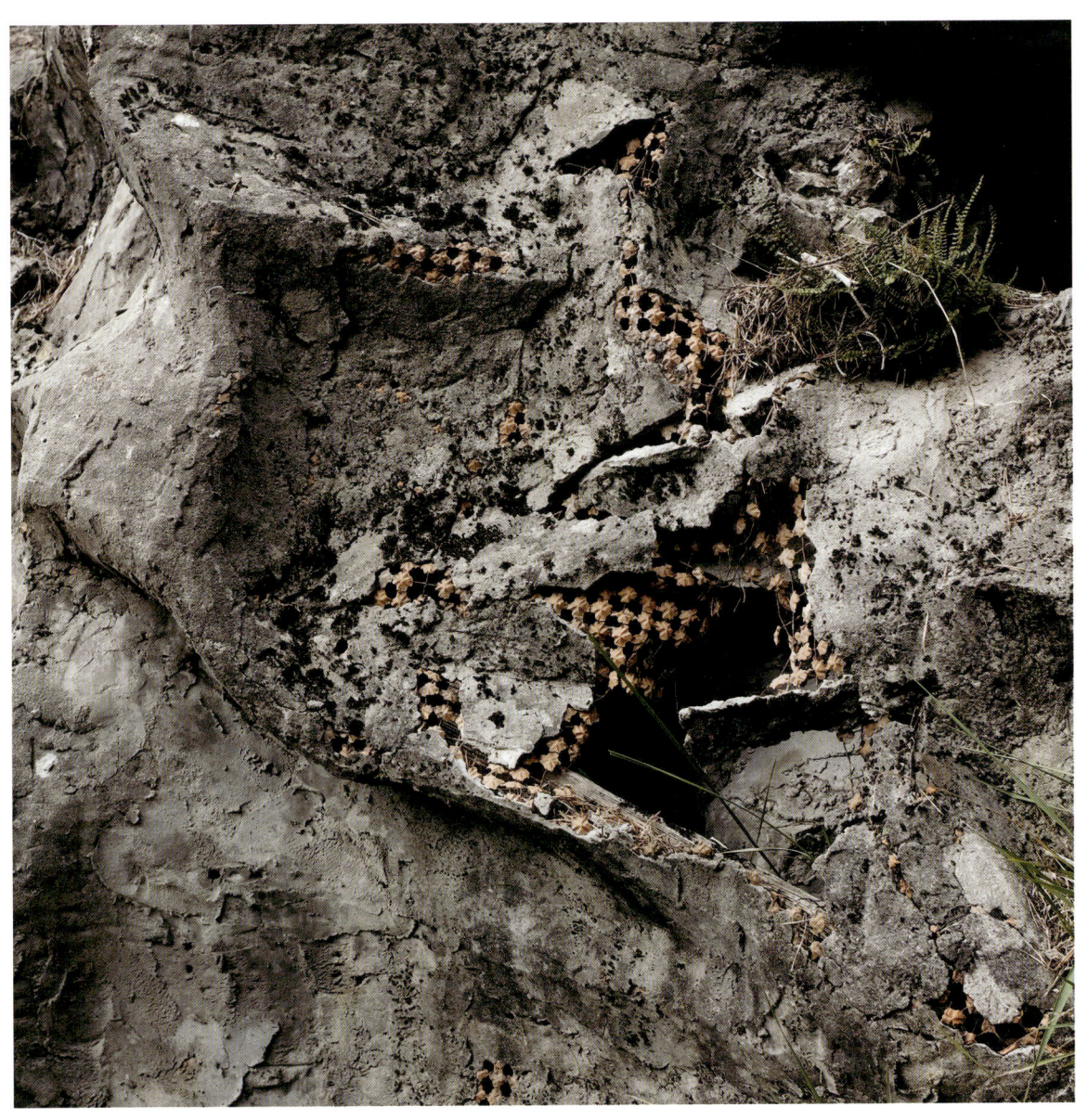

Vallo Alpino
Viévola

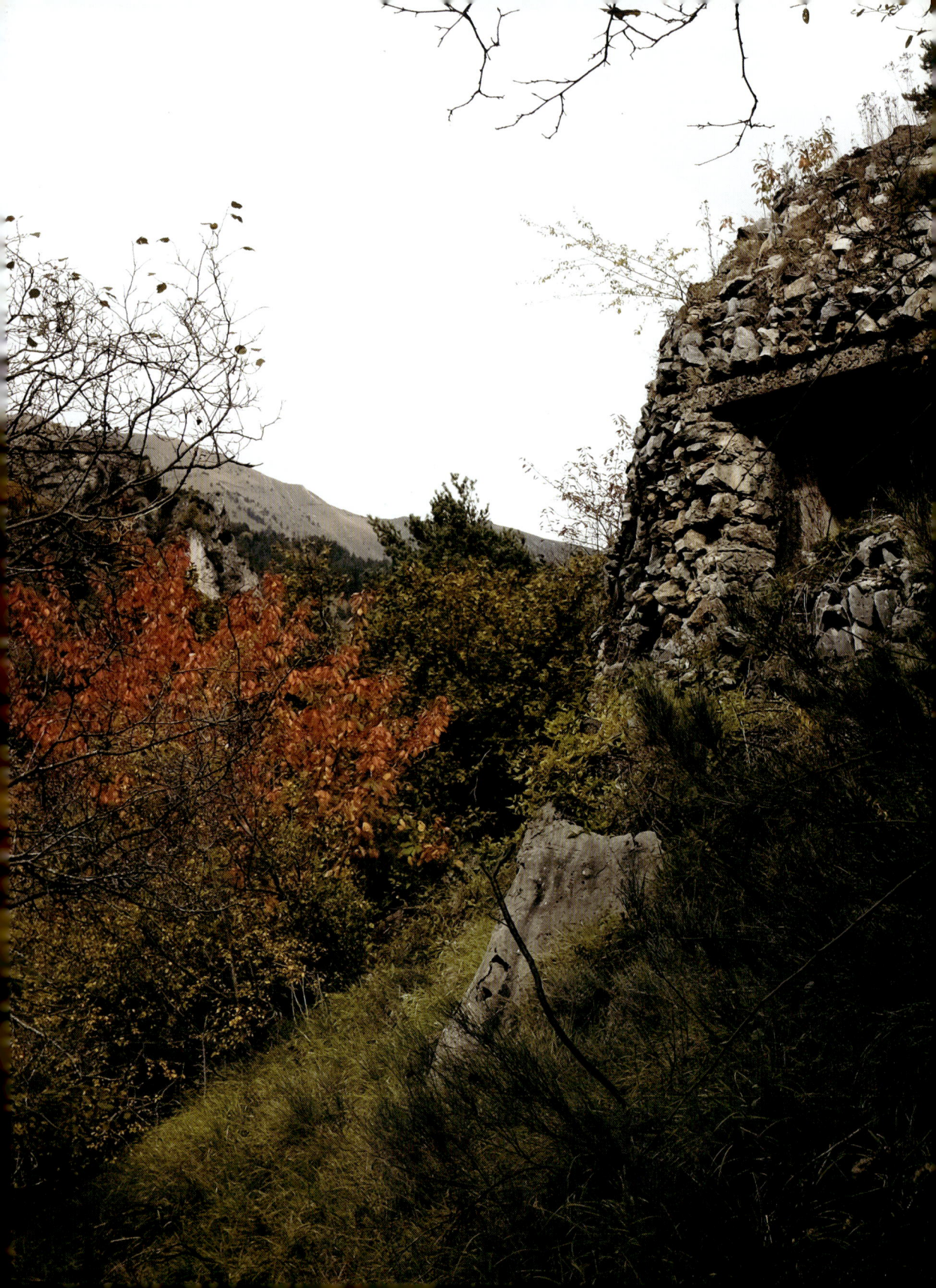

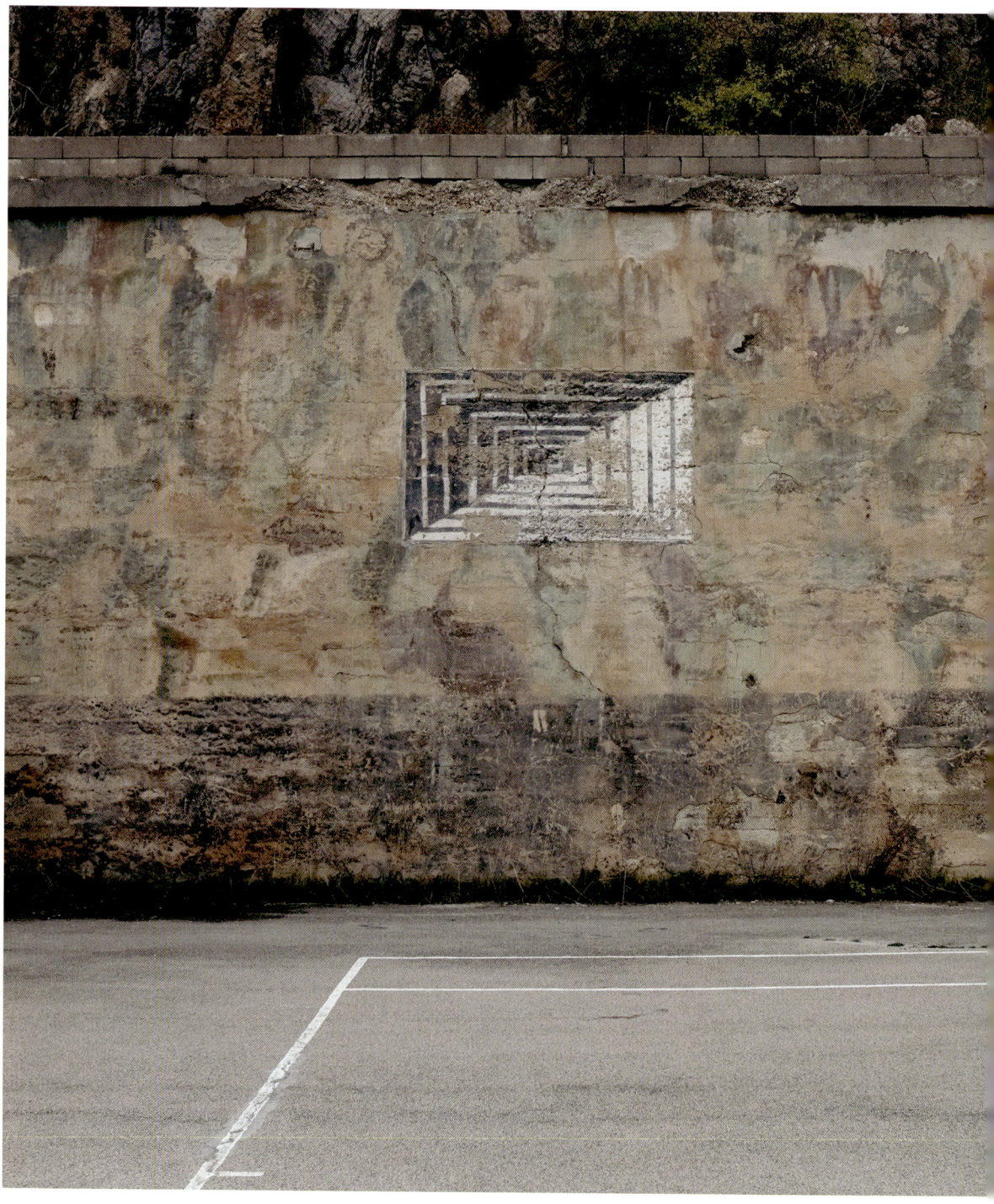

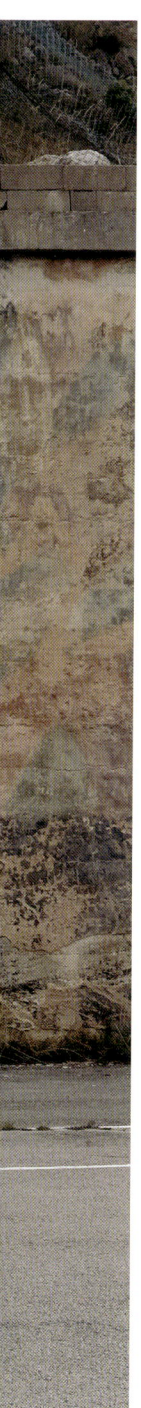

Left and below
Municipal tennis court, Rimplas

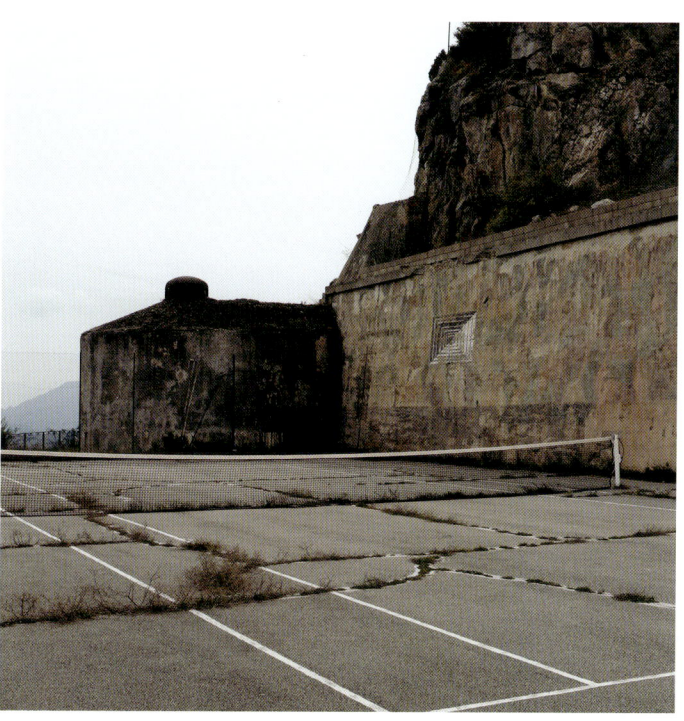

Following pages
Fort Central, Tende

The fort complexes at almost 2,000m above the Col de Tende were built by the Italians in the 19th century, and in their prime housed more than 1,000 people. They were added to for the *Vallo Alpino* before the Second World War.

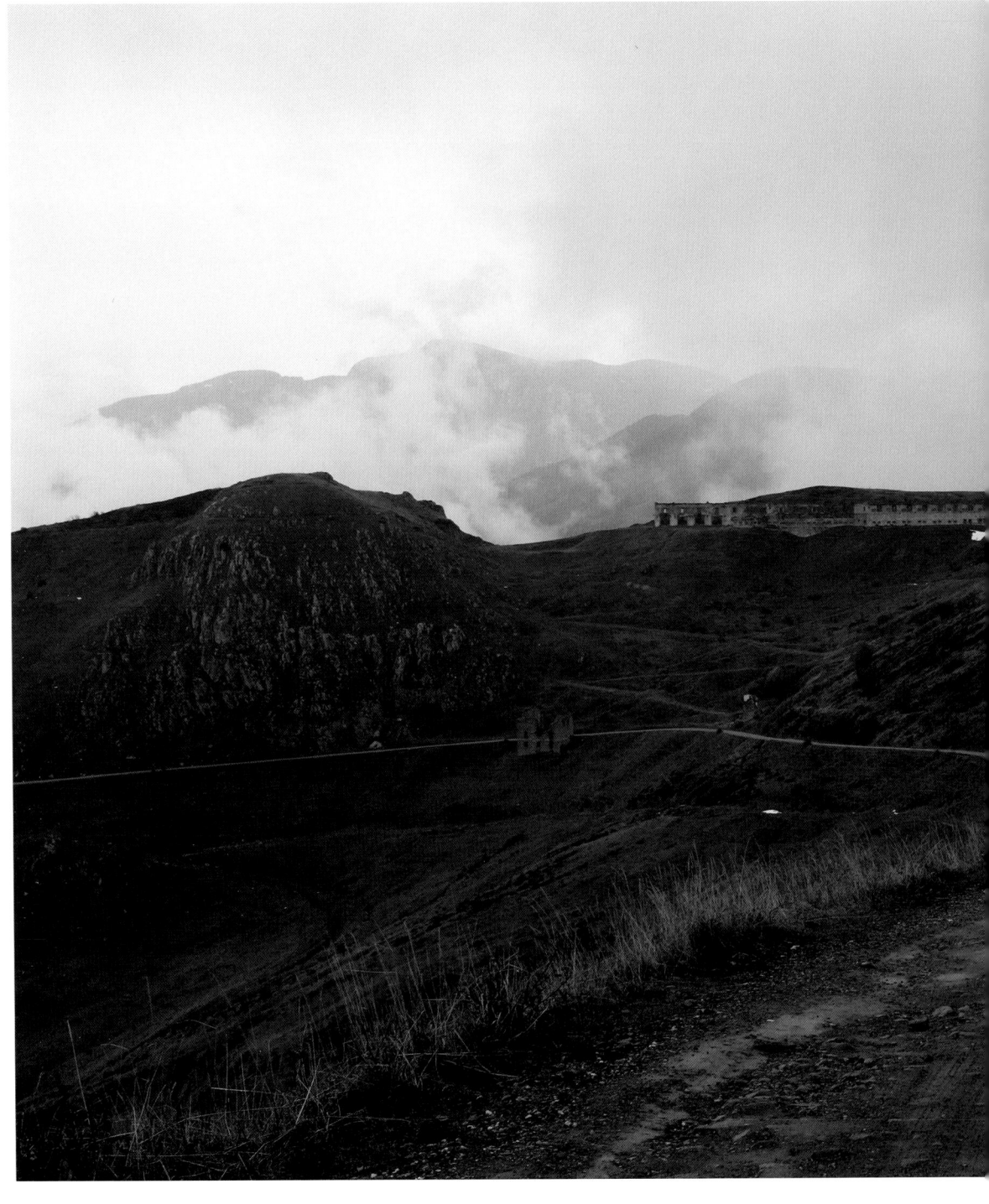

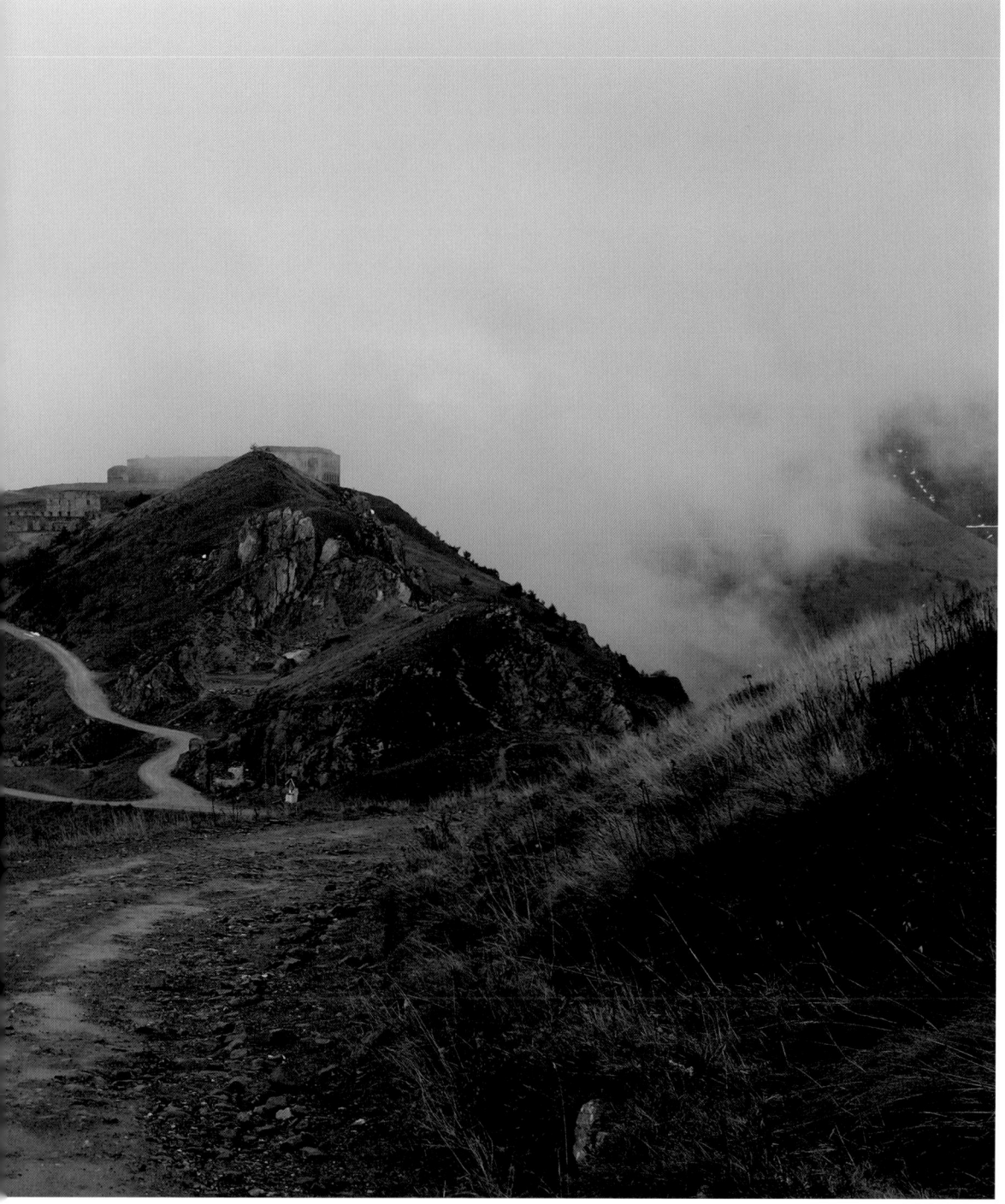

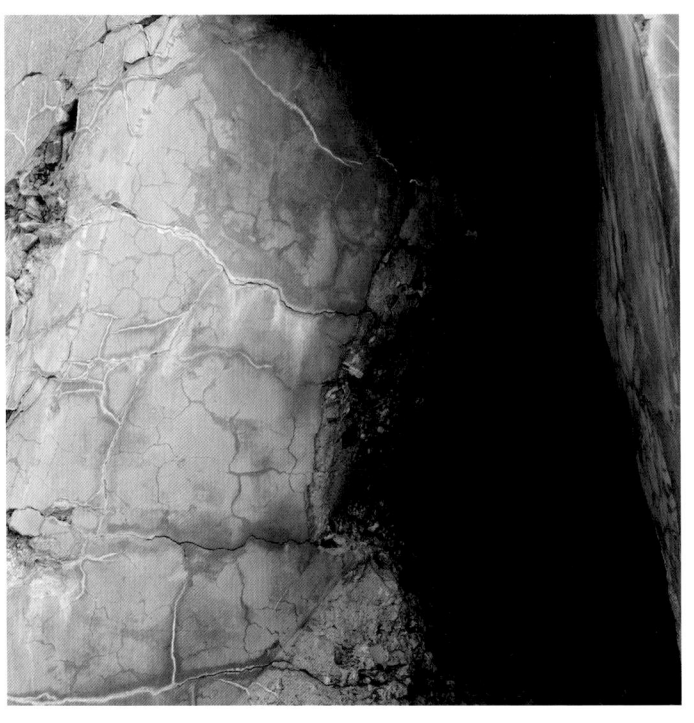

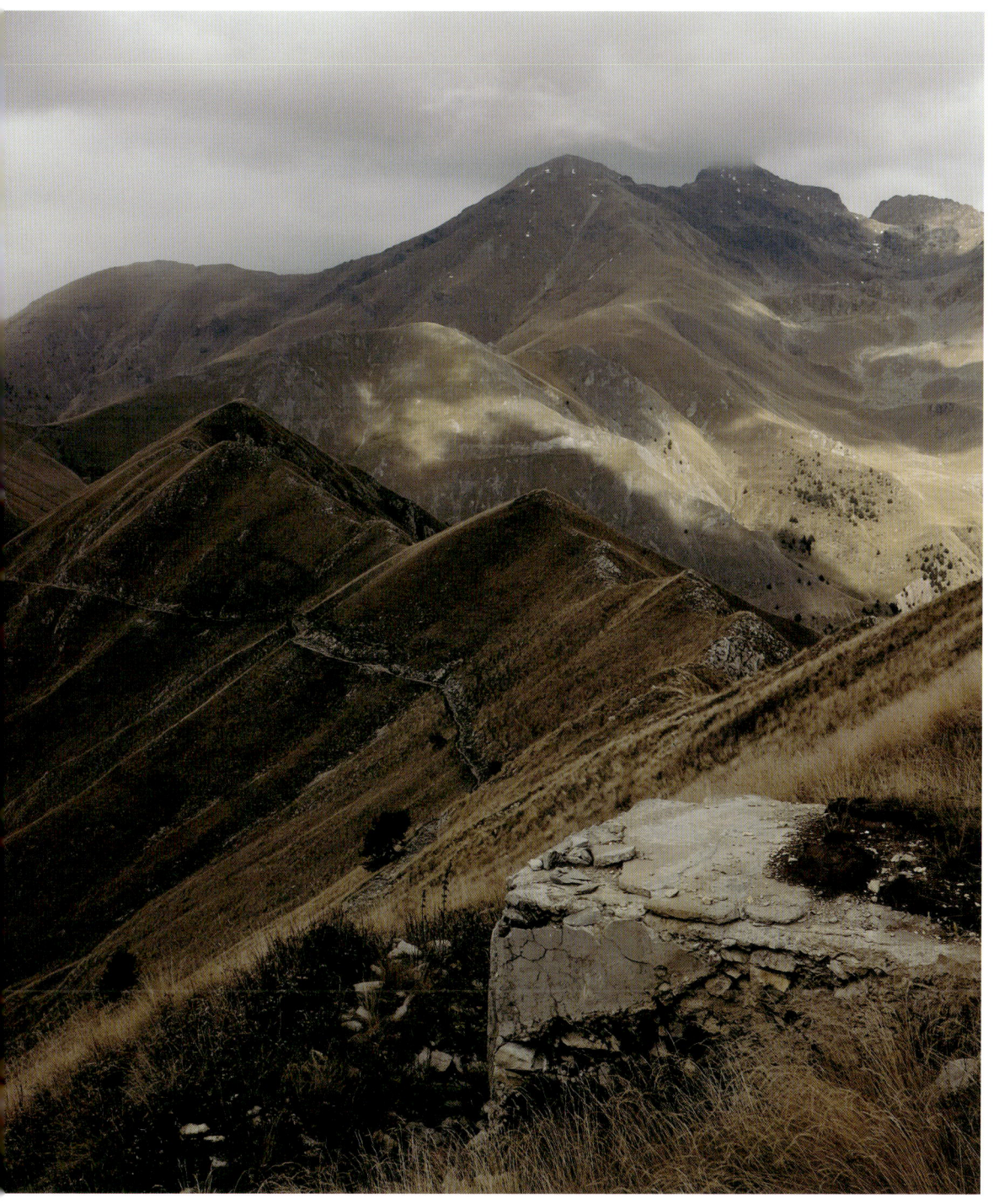

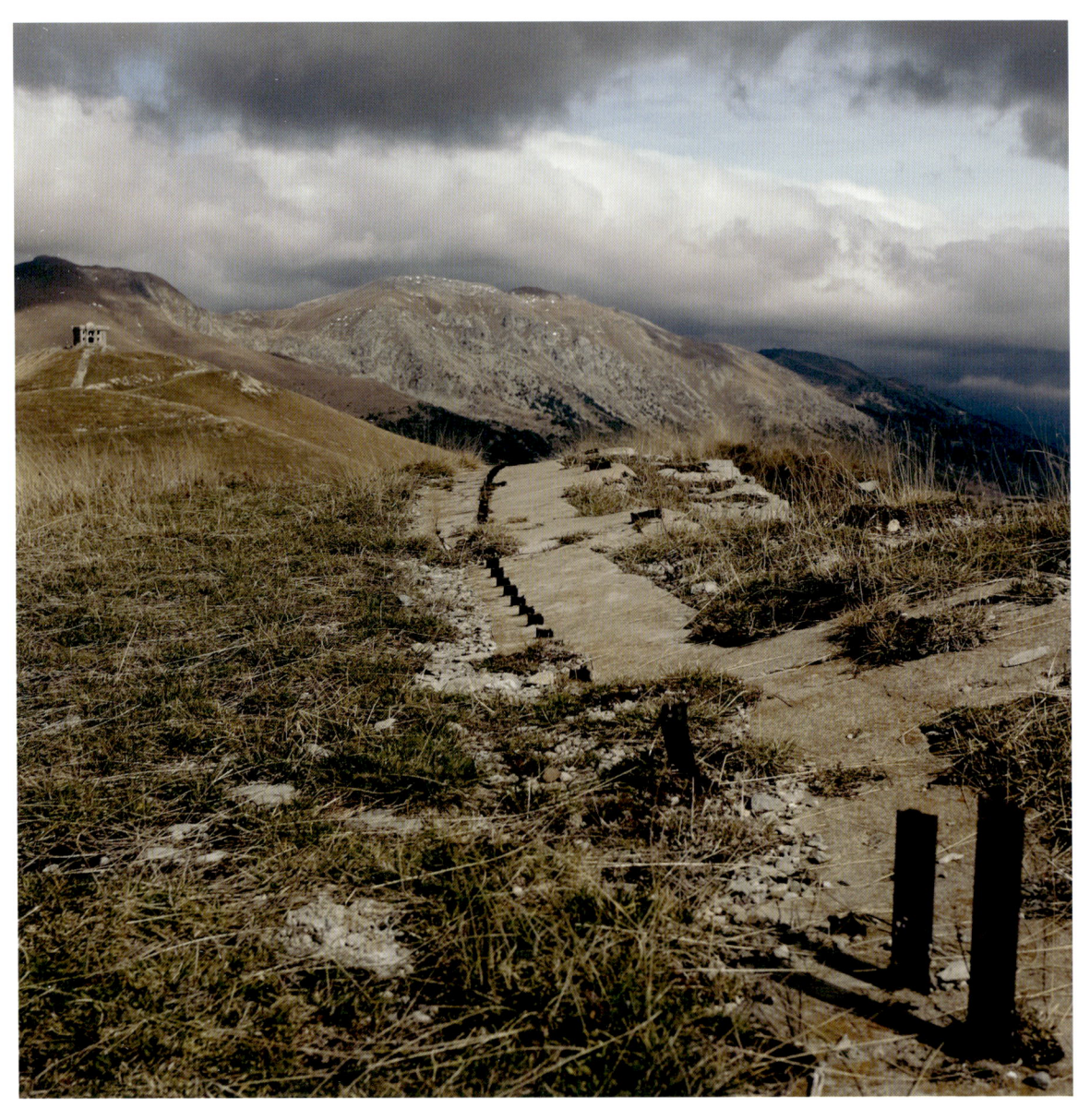

Above and opposite
Redoute de la Forca
SFAM / Authion

Following pages
Petit ouvrage de la Moutière
SFD / Jausiers
One officer, 42 men

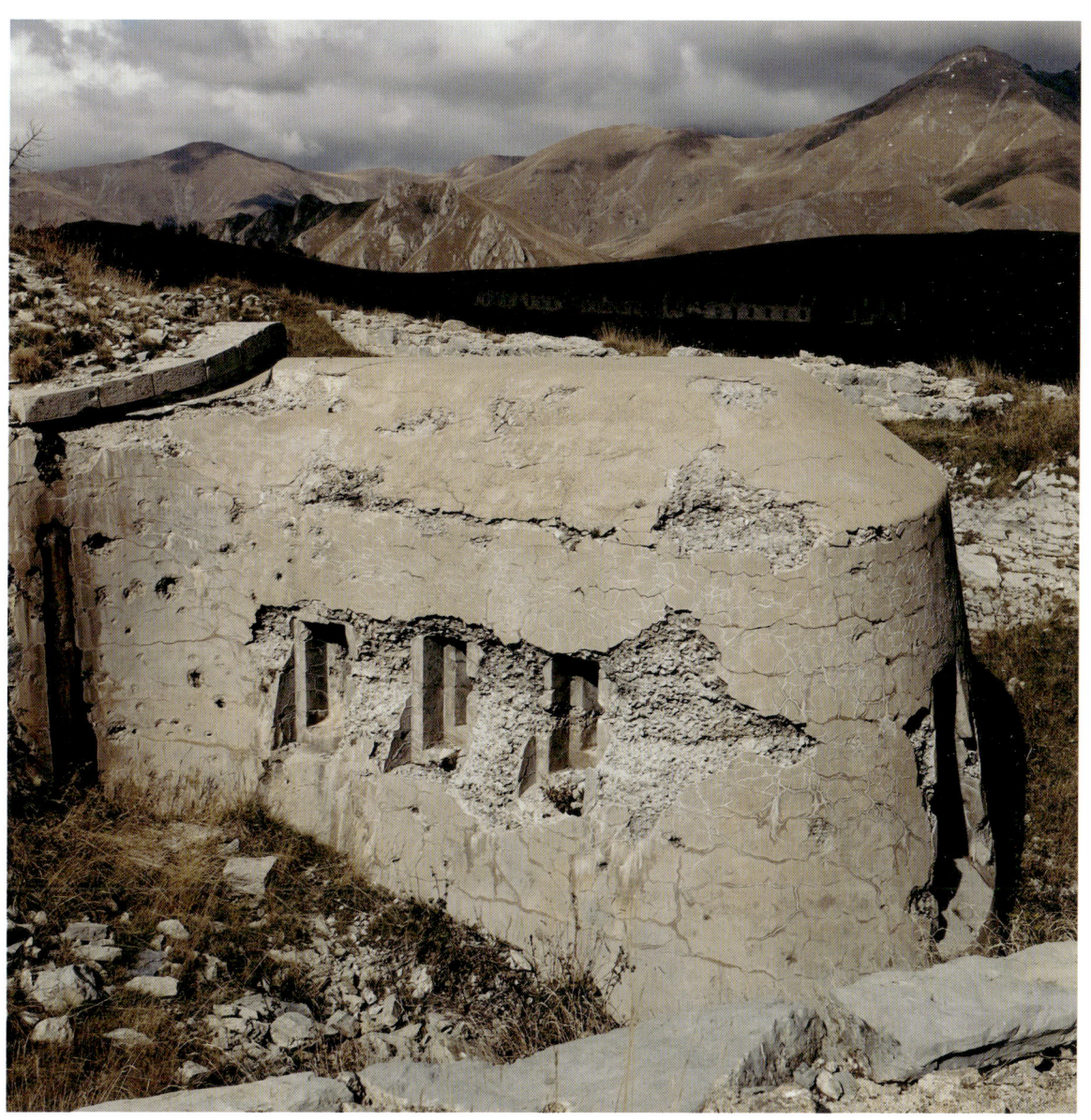

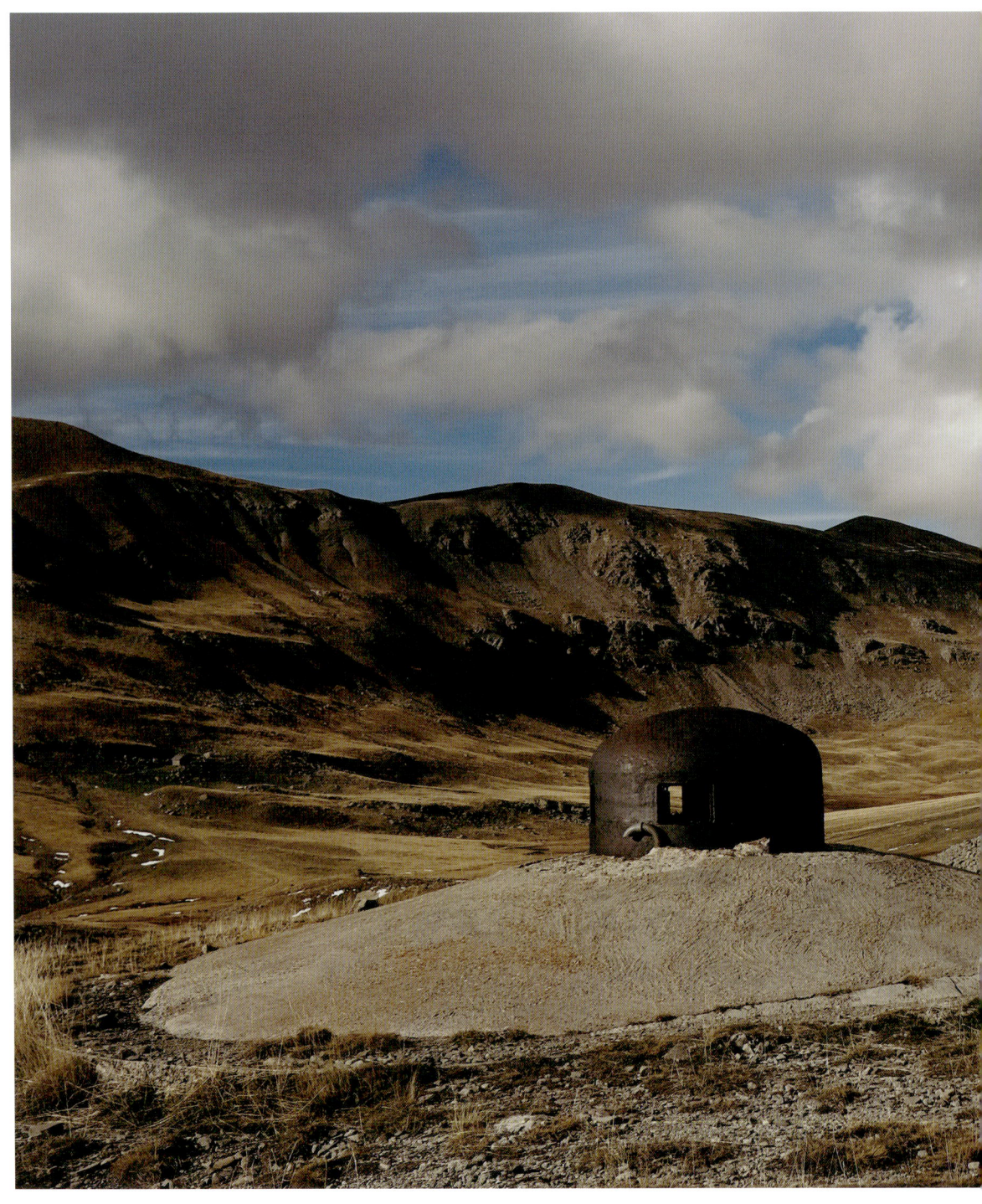

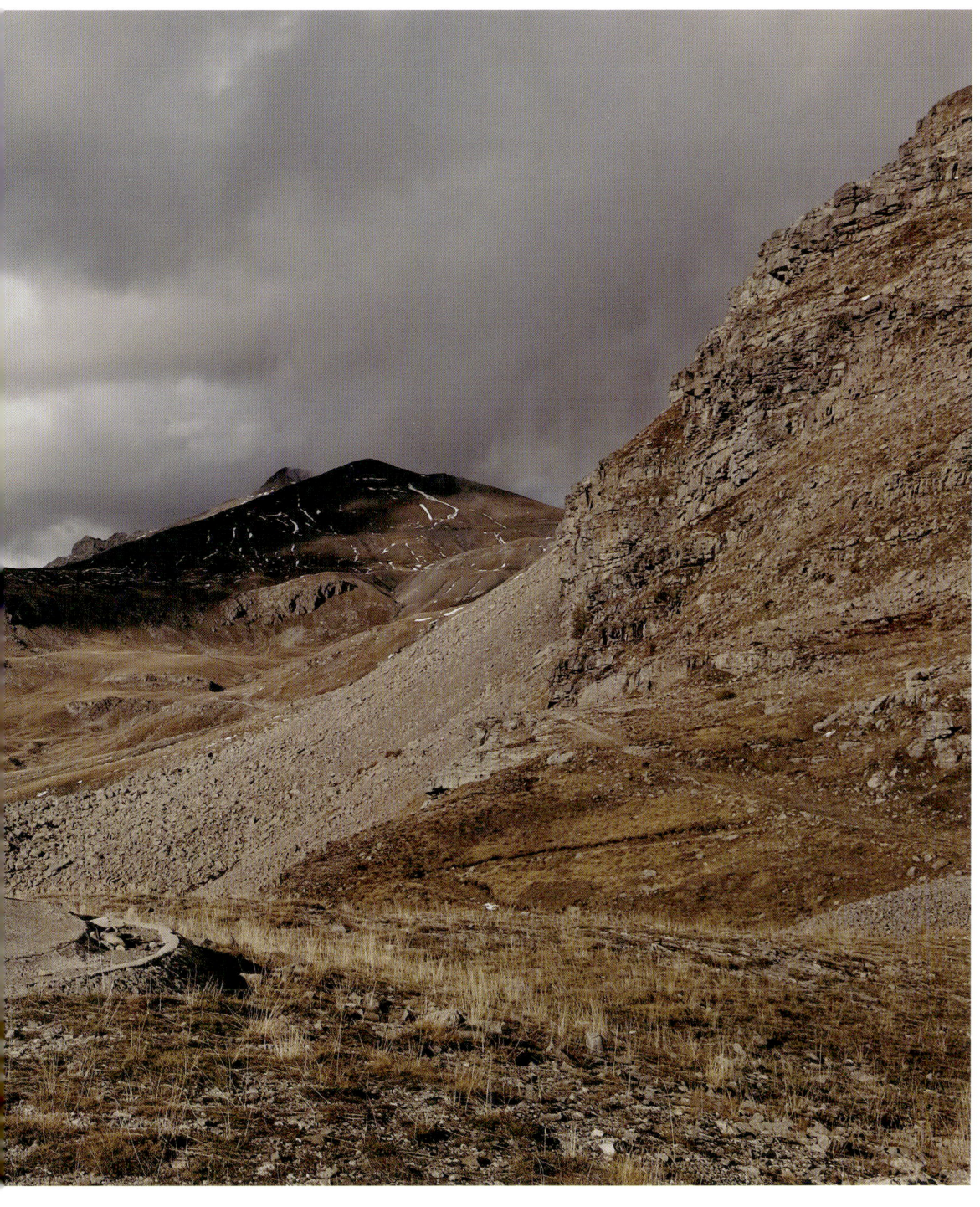

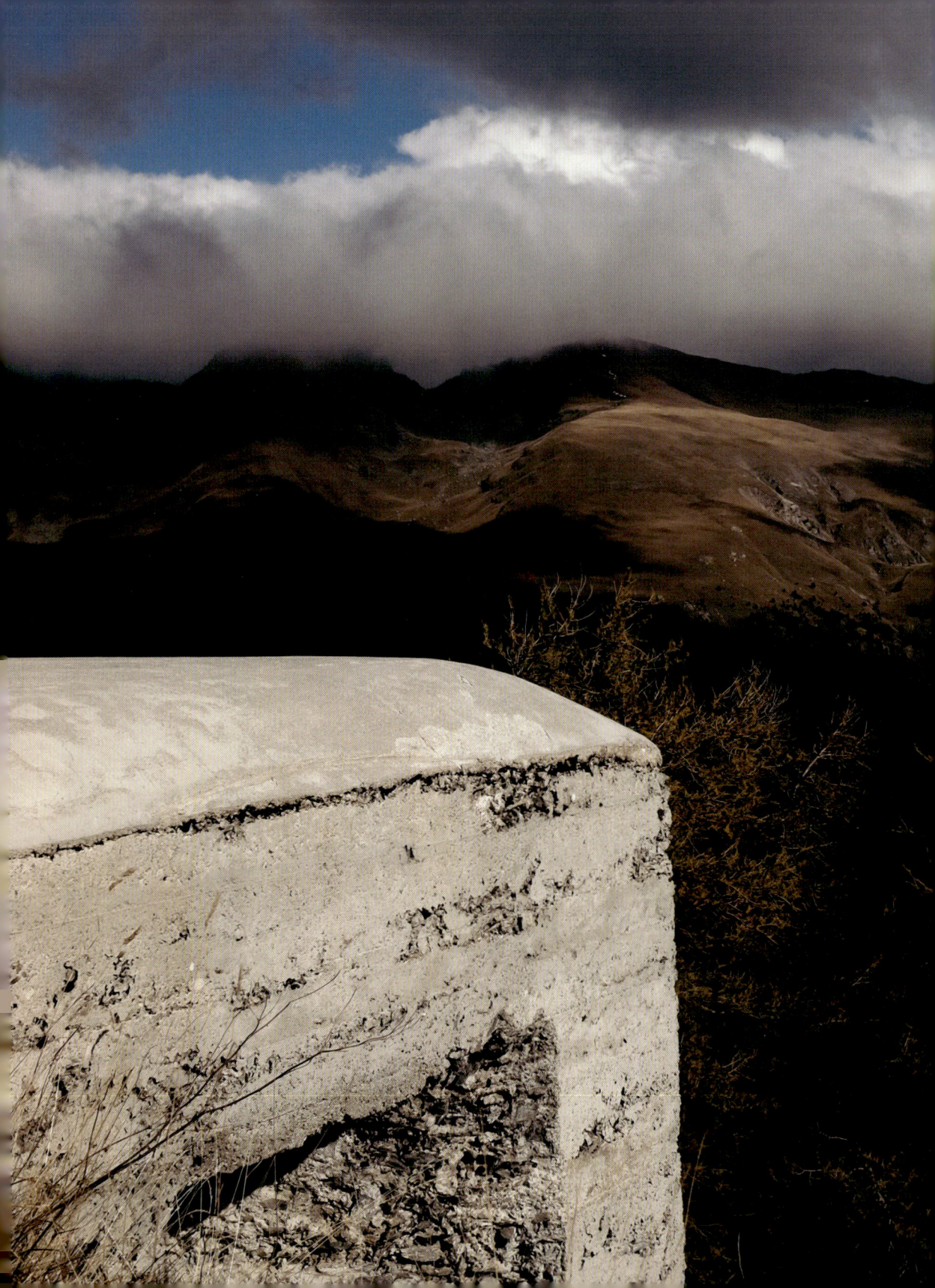

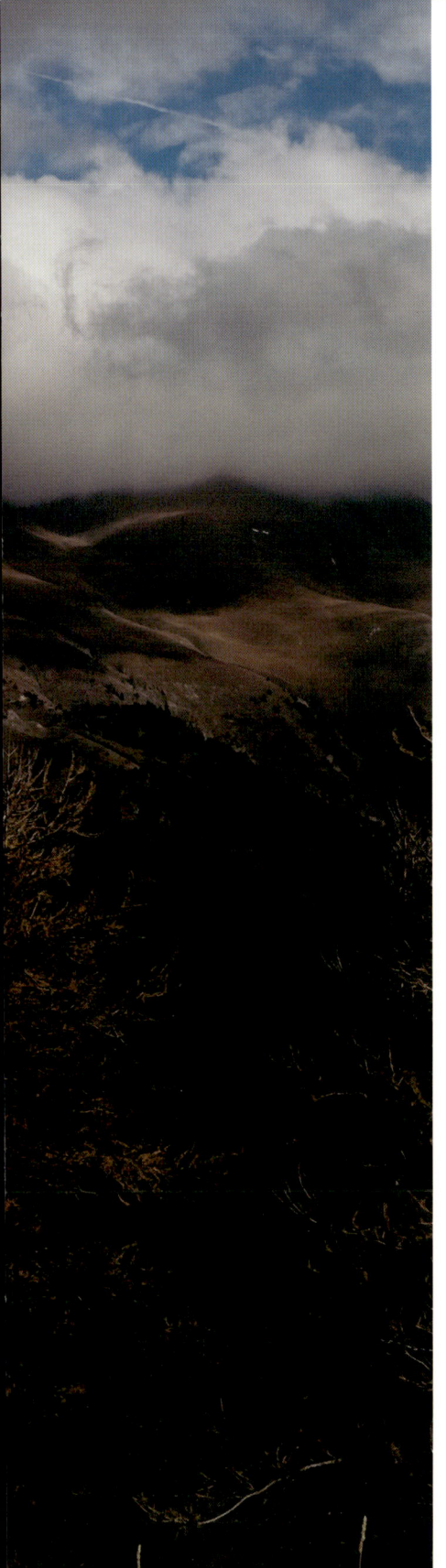

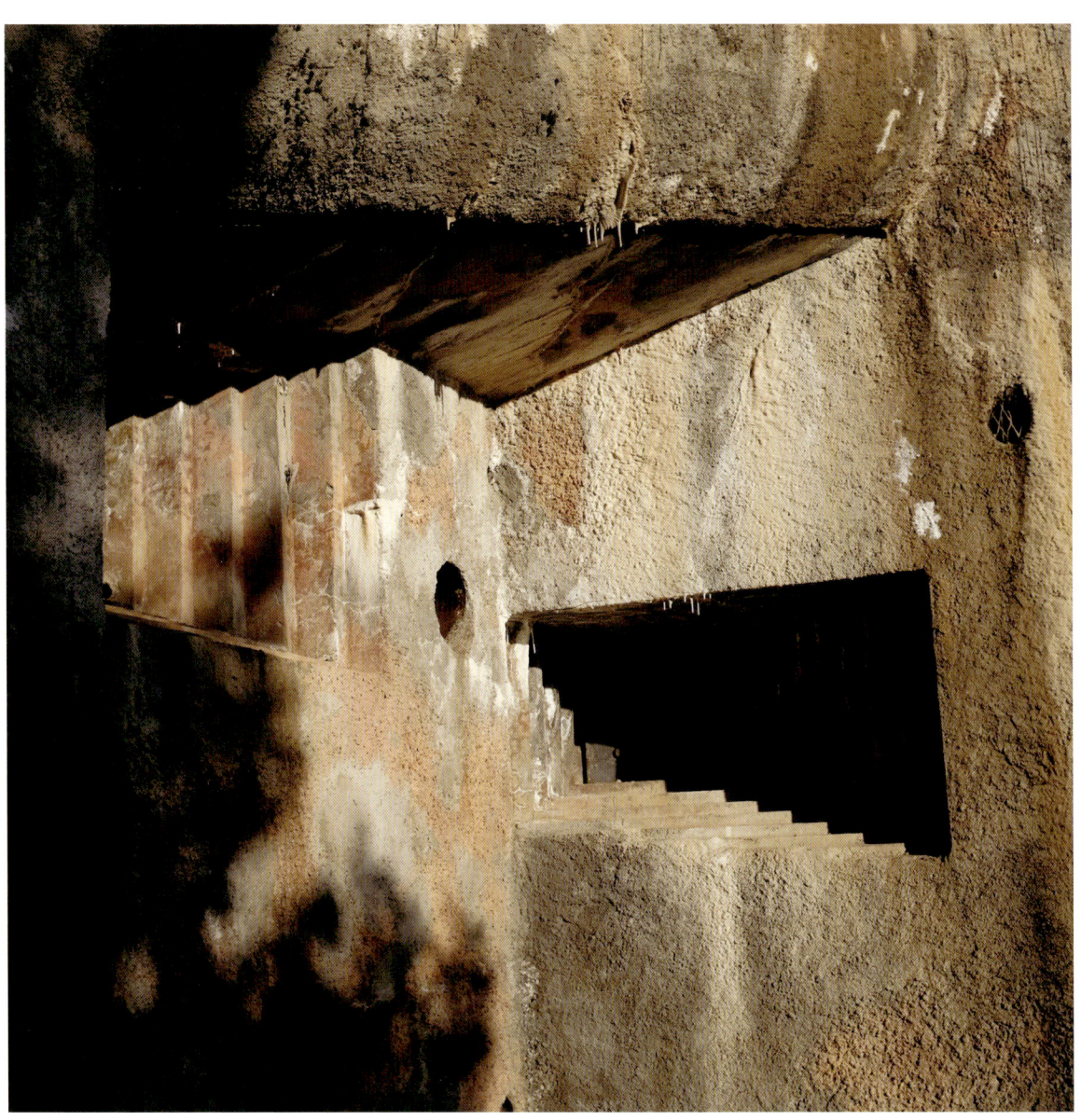

Following pages
Avant-poste du Col des Fourches
SFD / Jausiers
One officer, 42 men

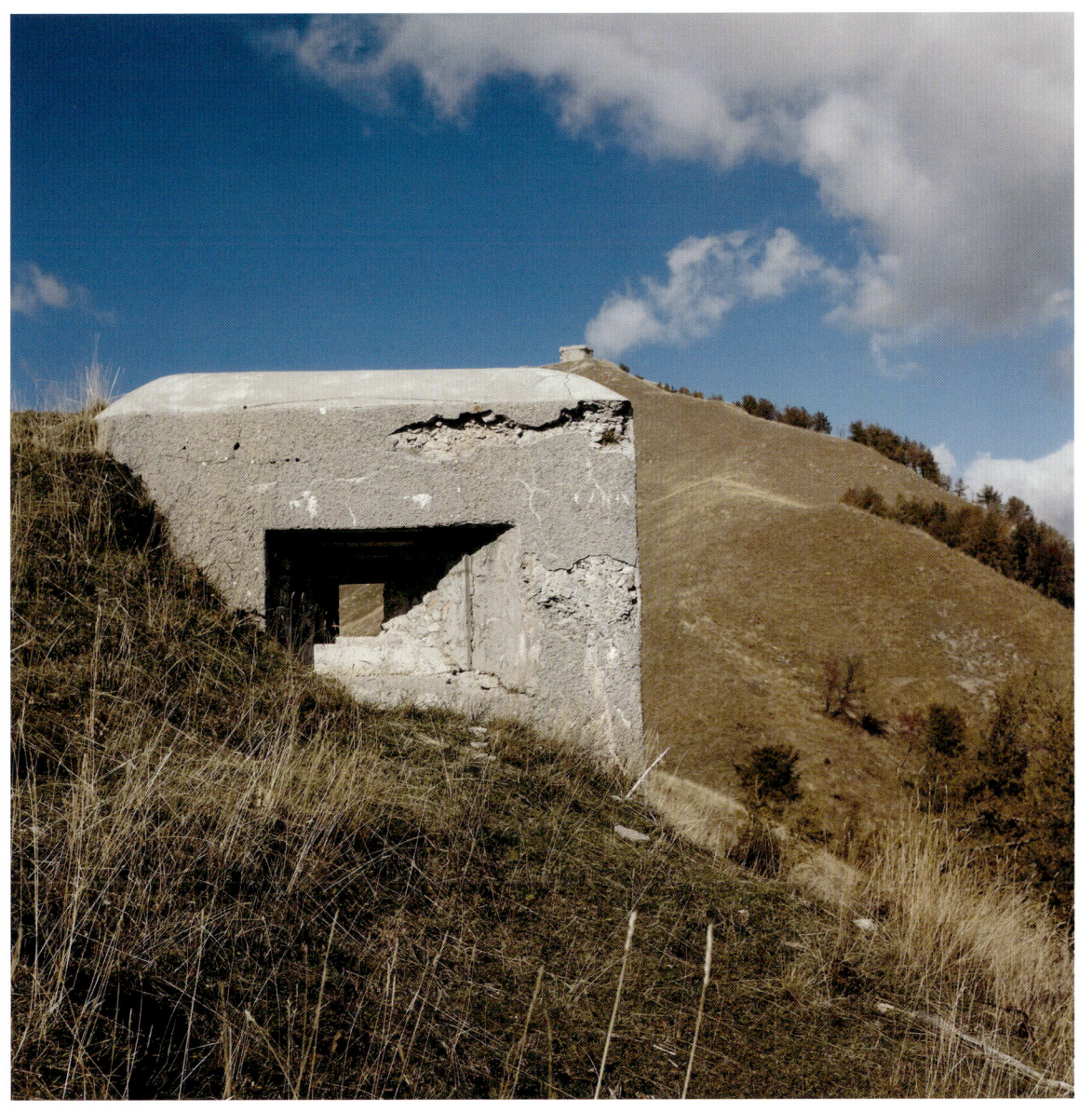

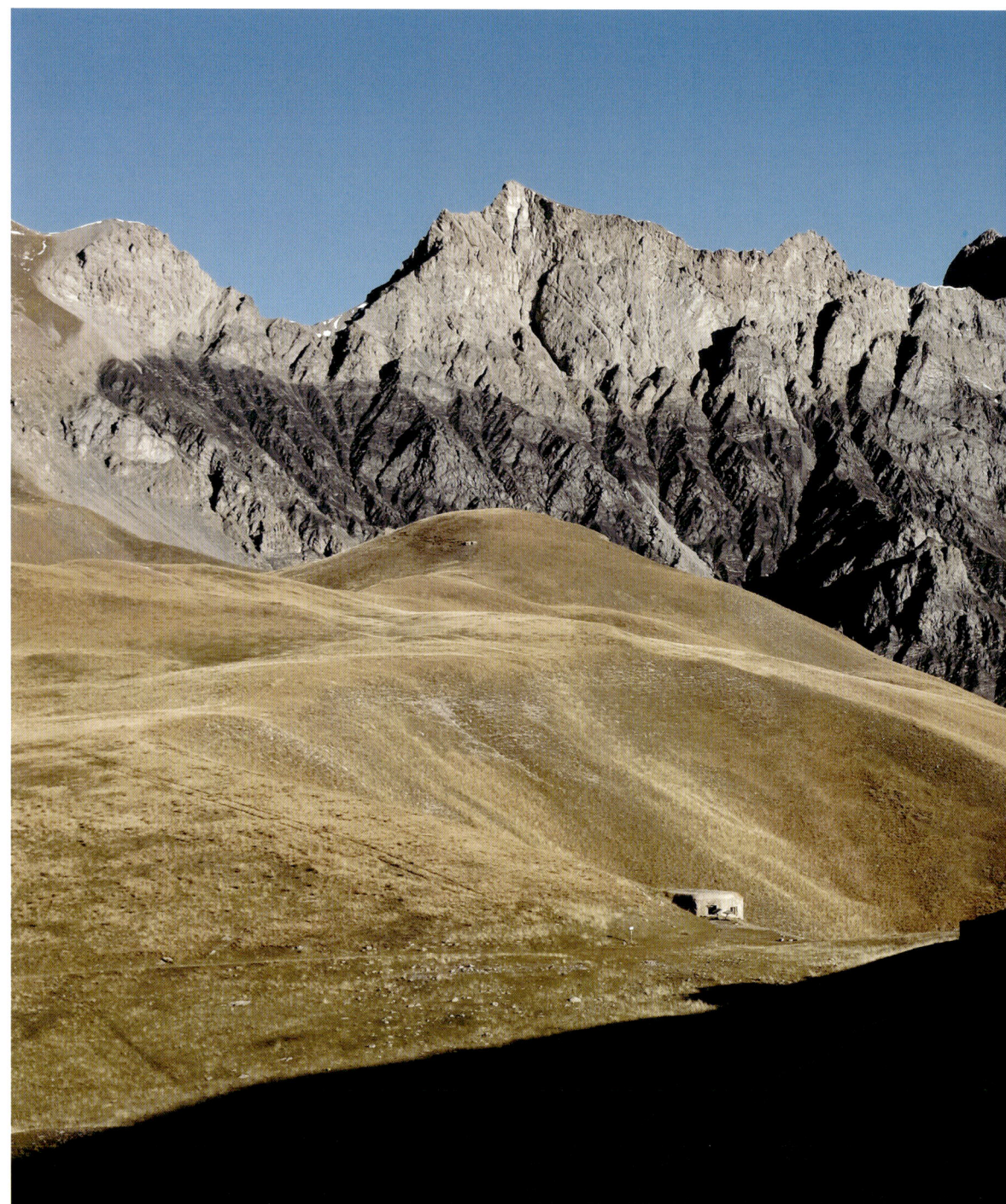

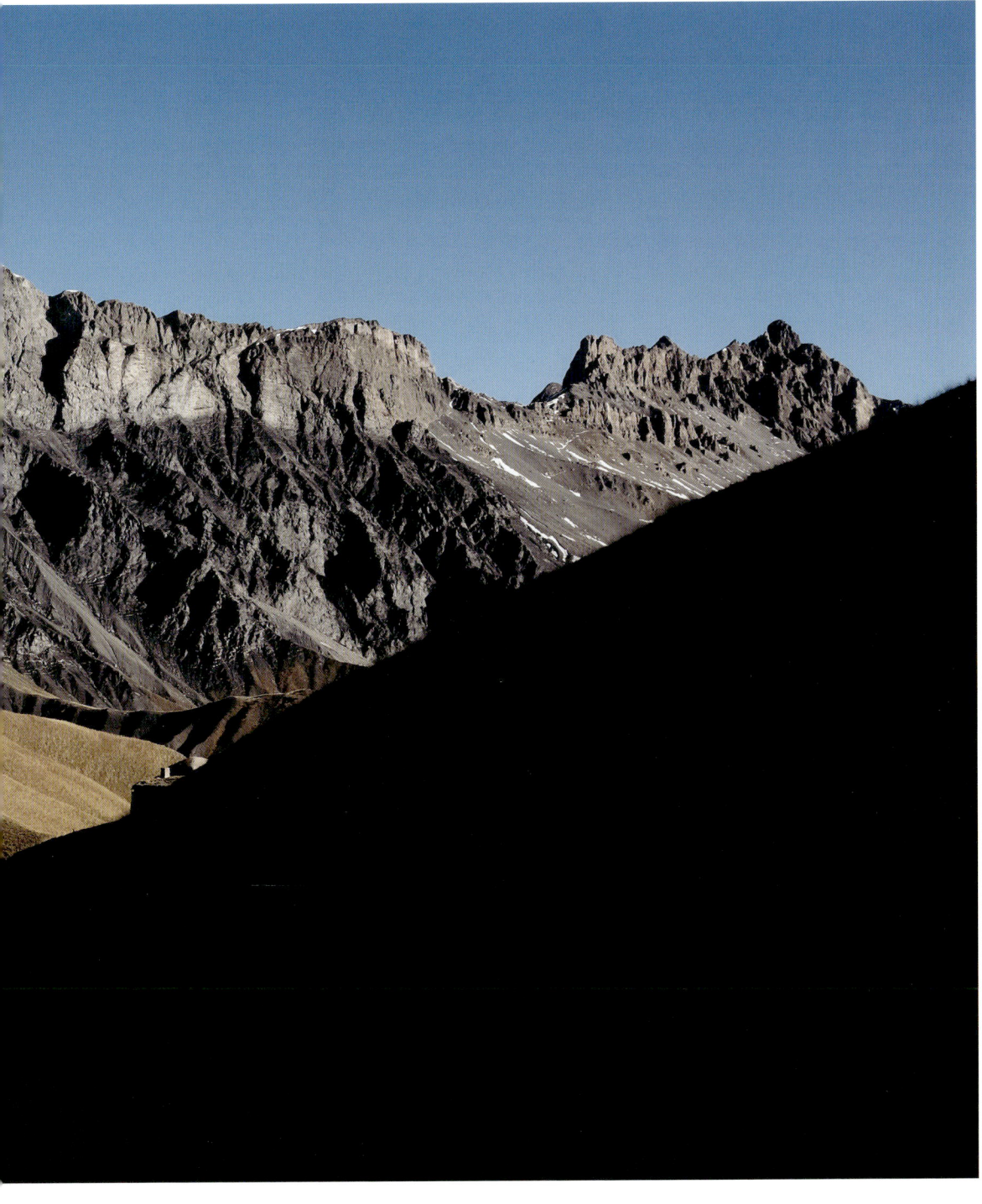

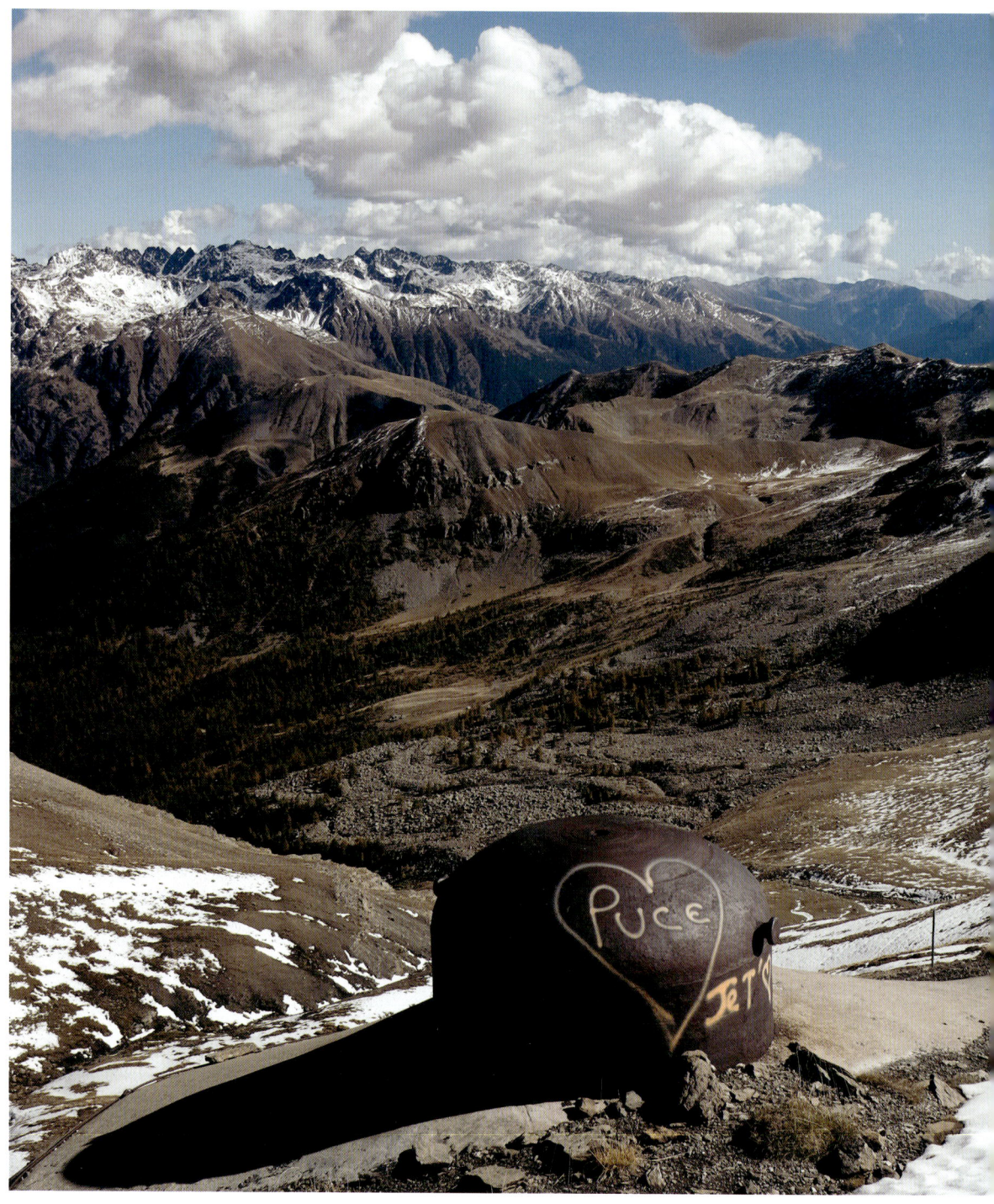

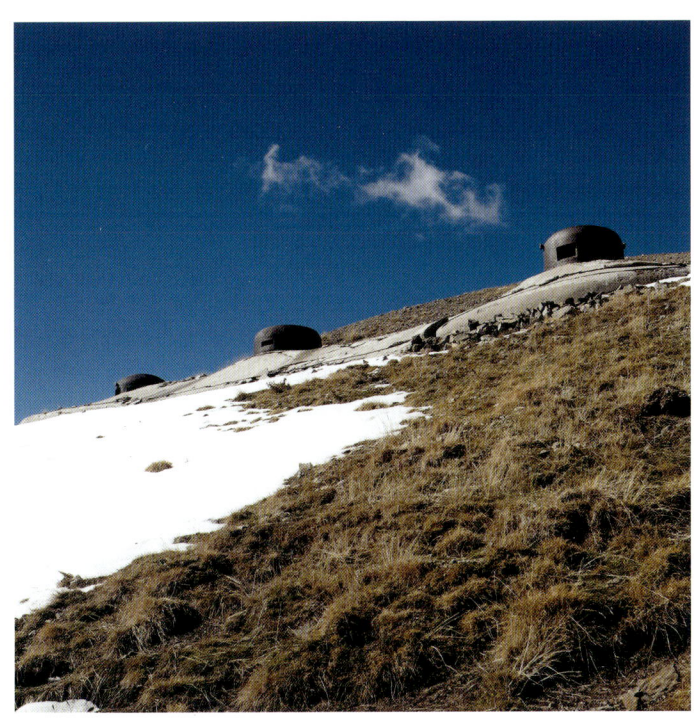

Petit ouvrage des Granges Communes
SFD / Jausiers
Two officers, 30 men

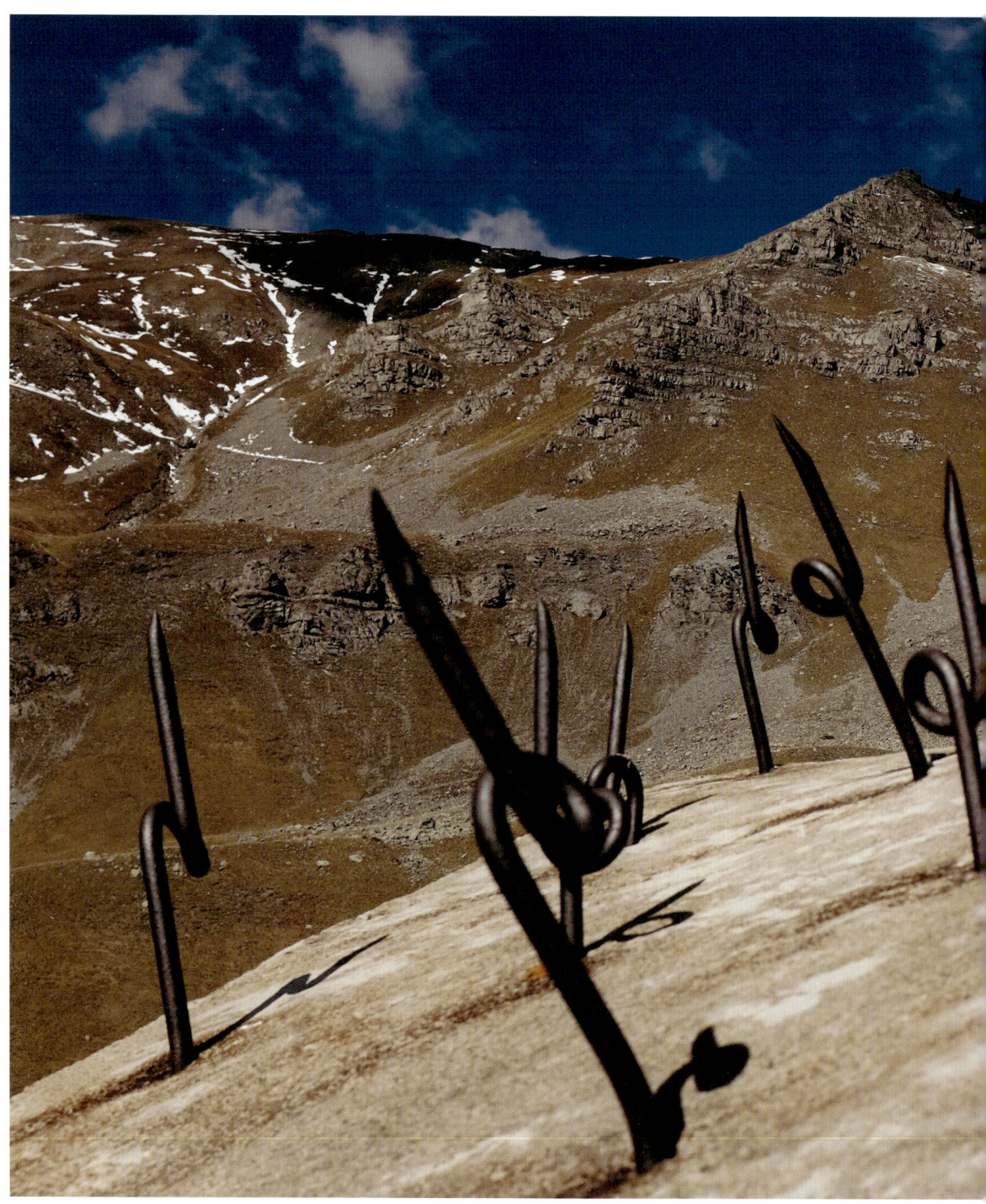

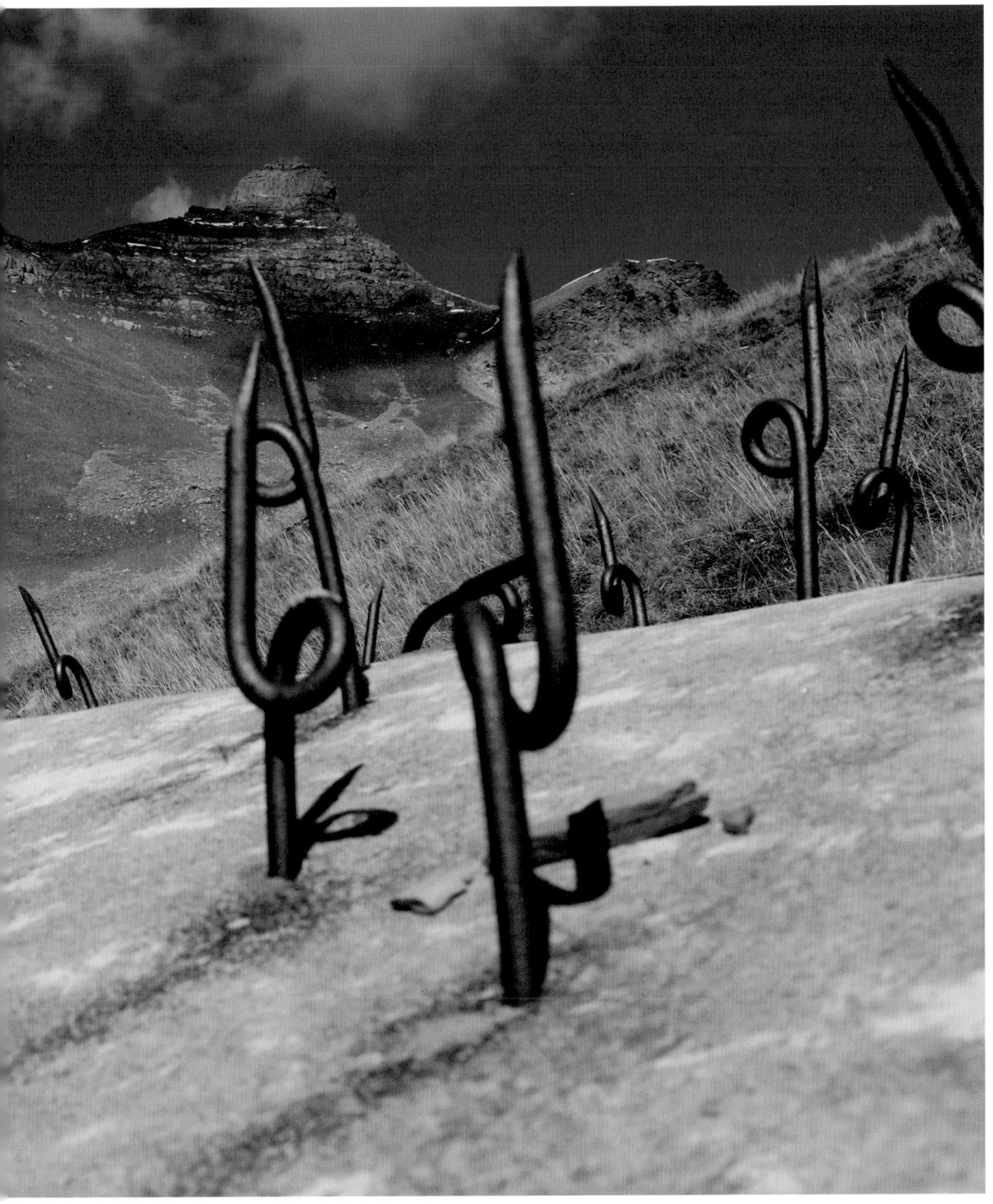

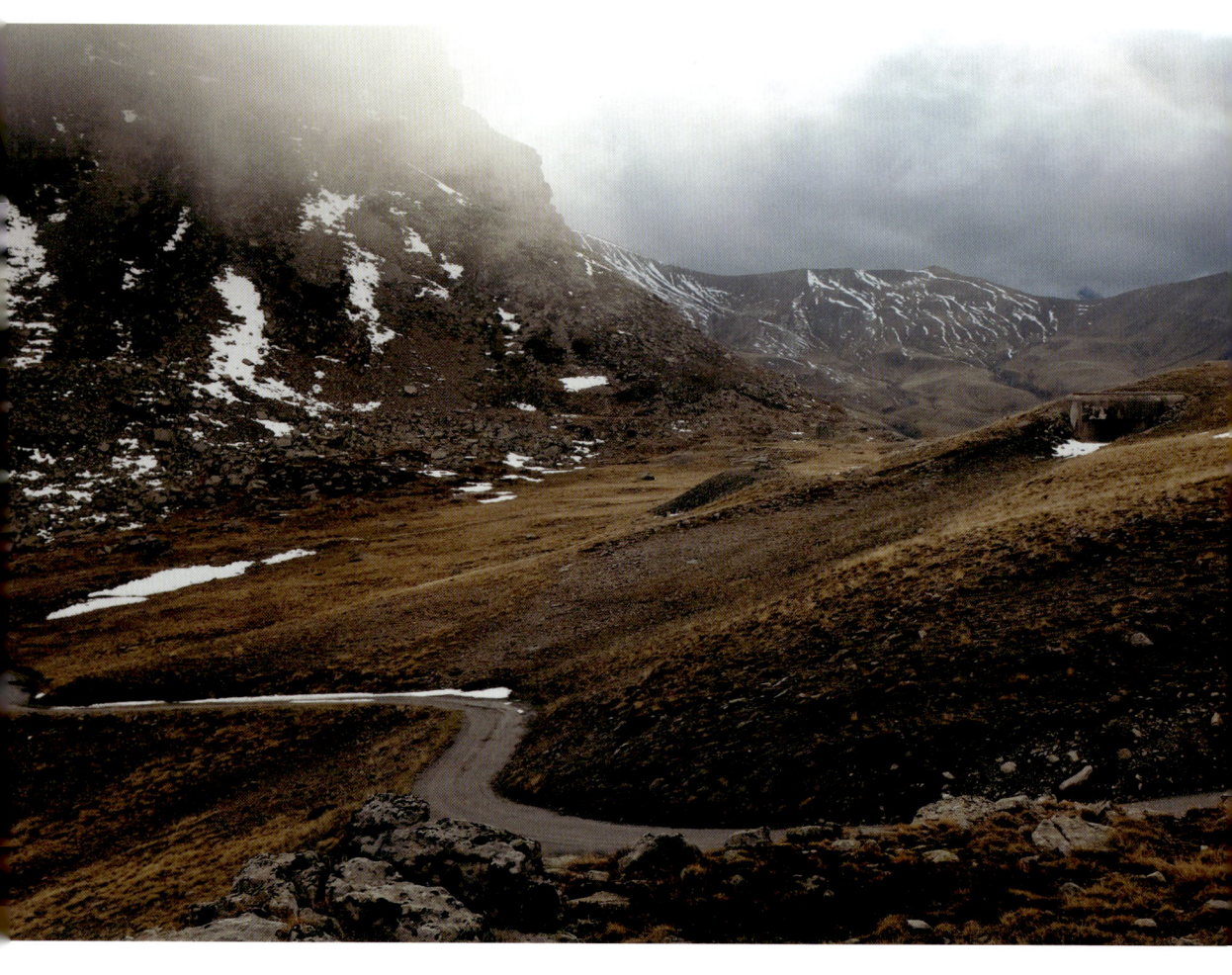

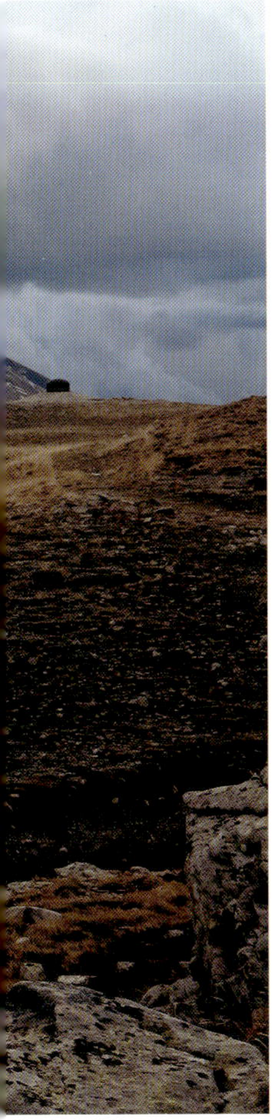

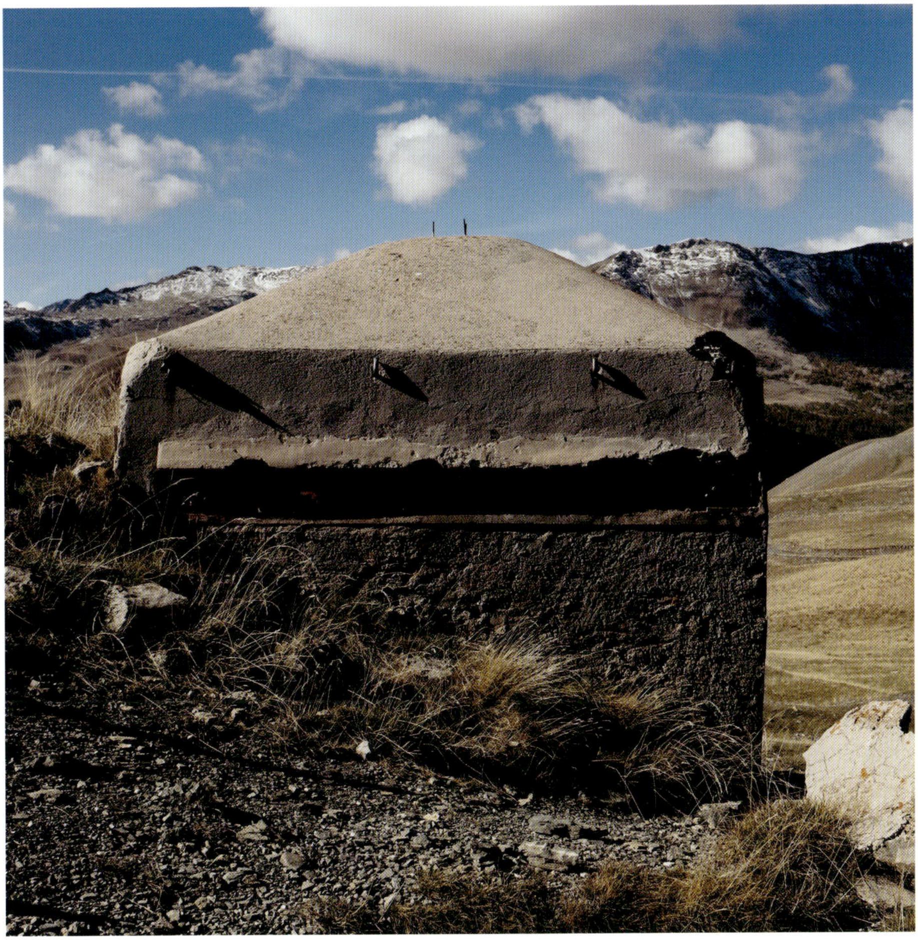

101

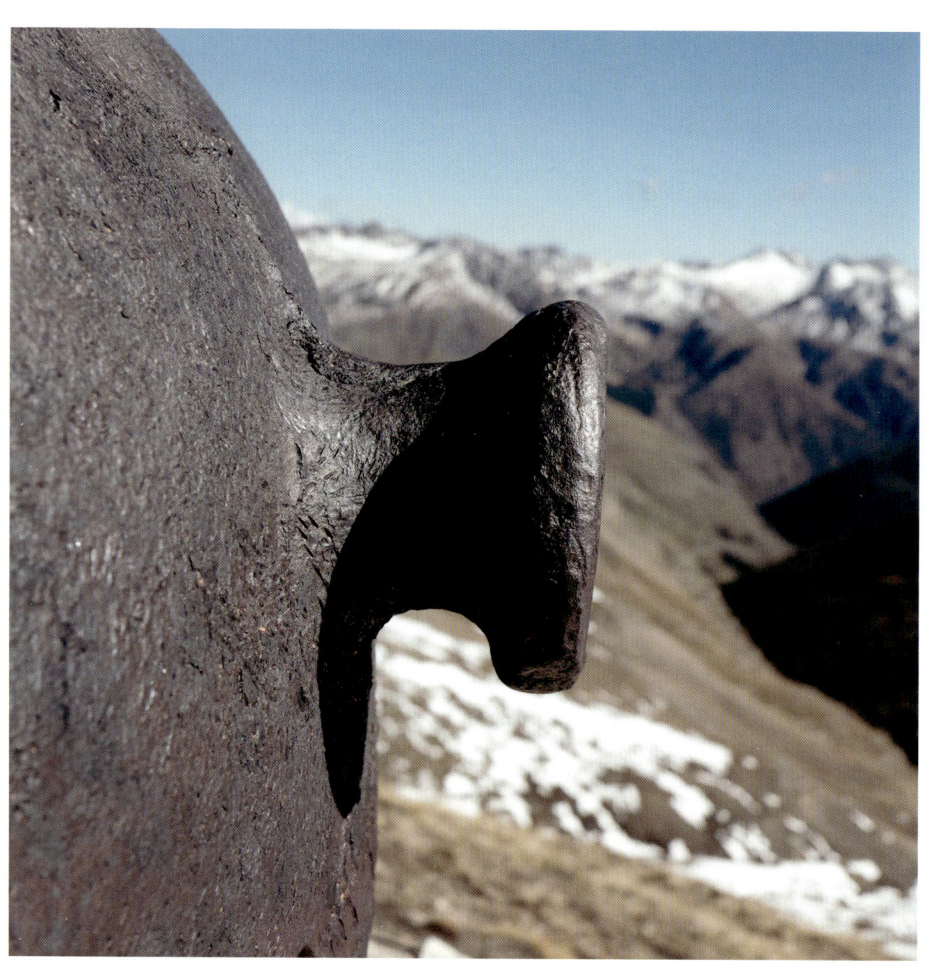

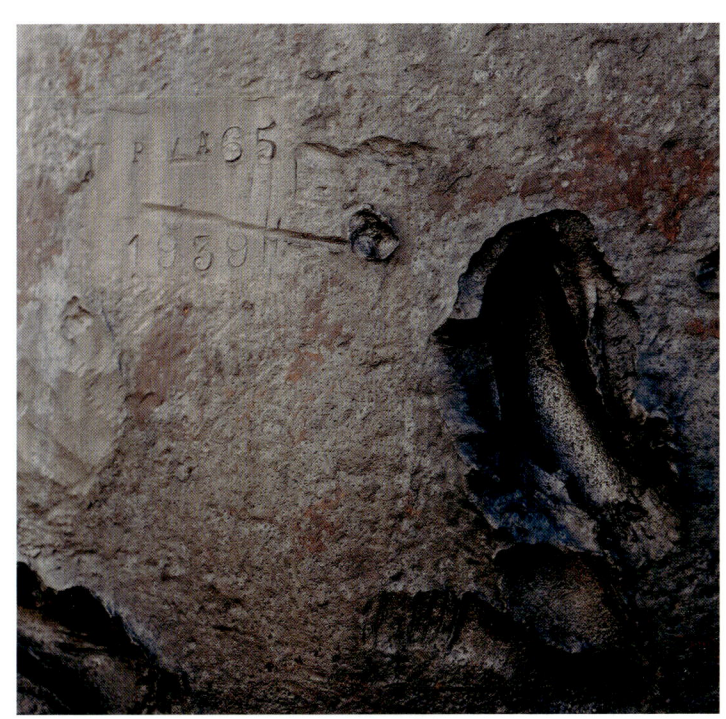
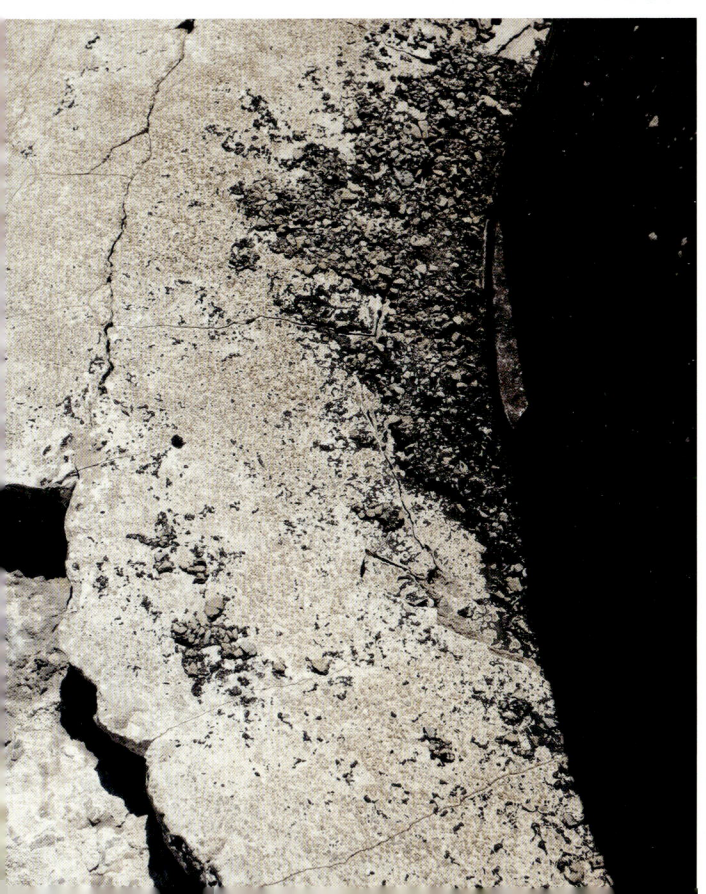

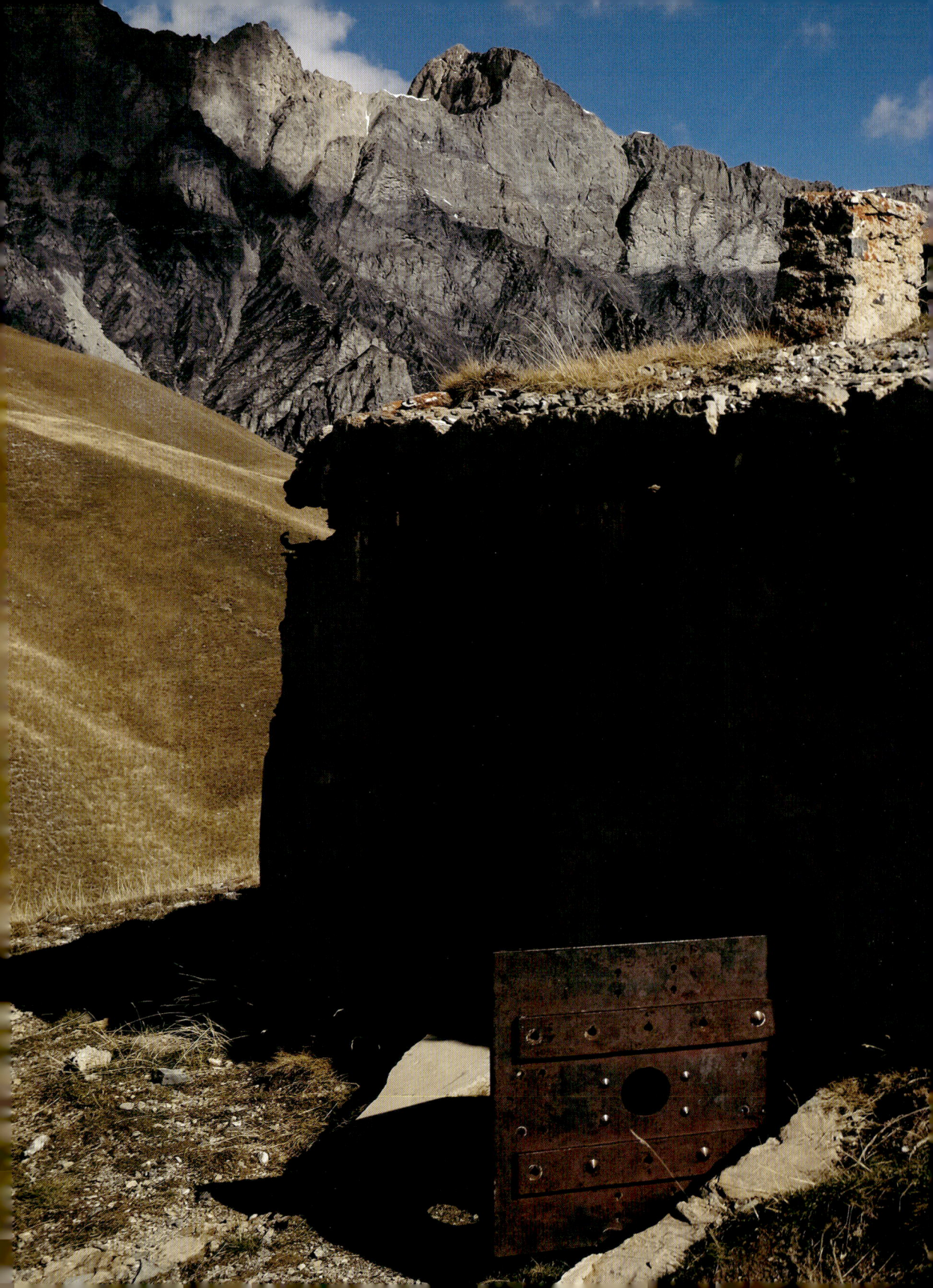

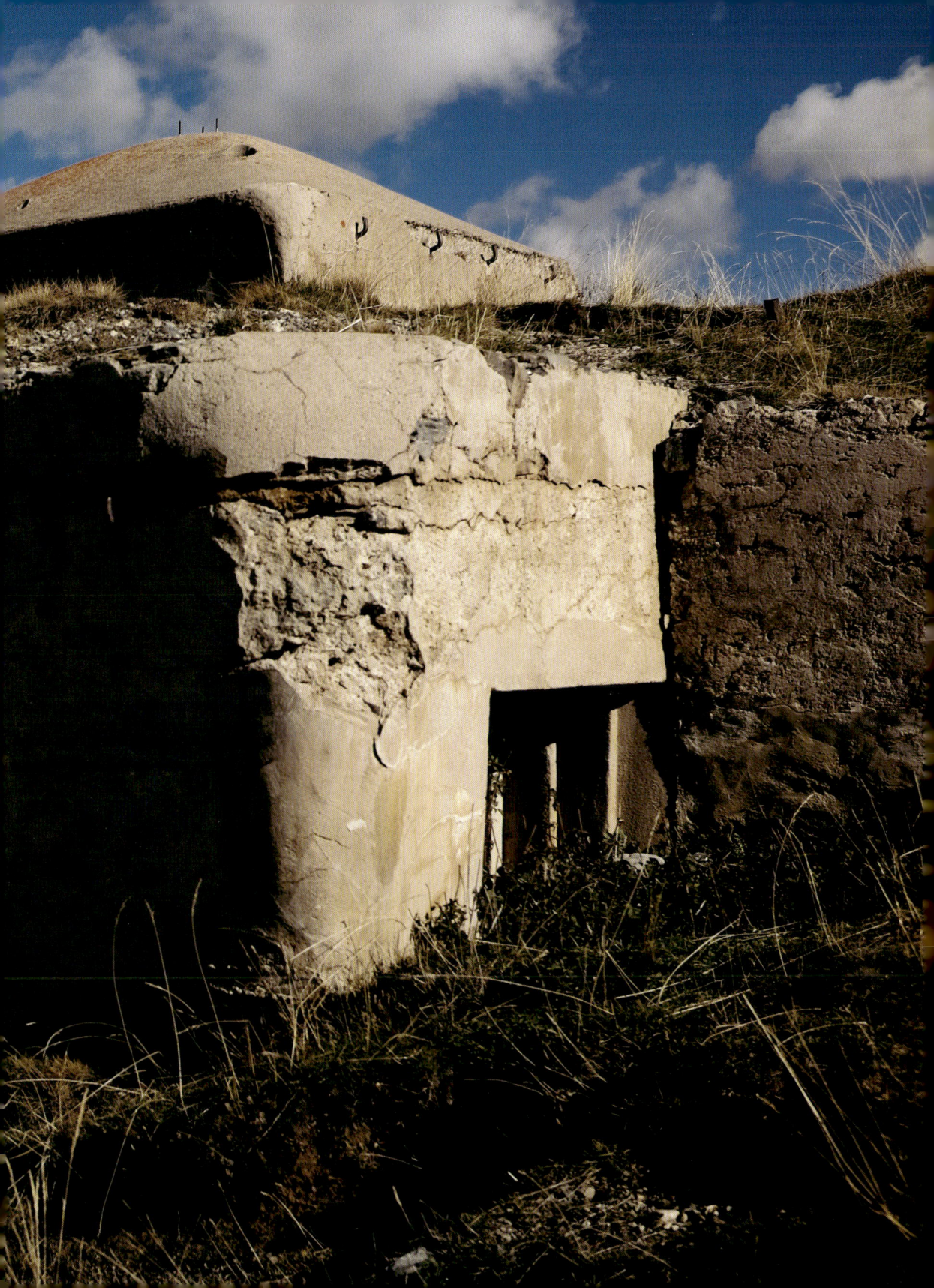

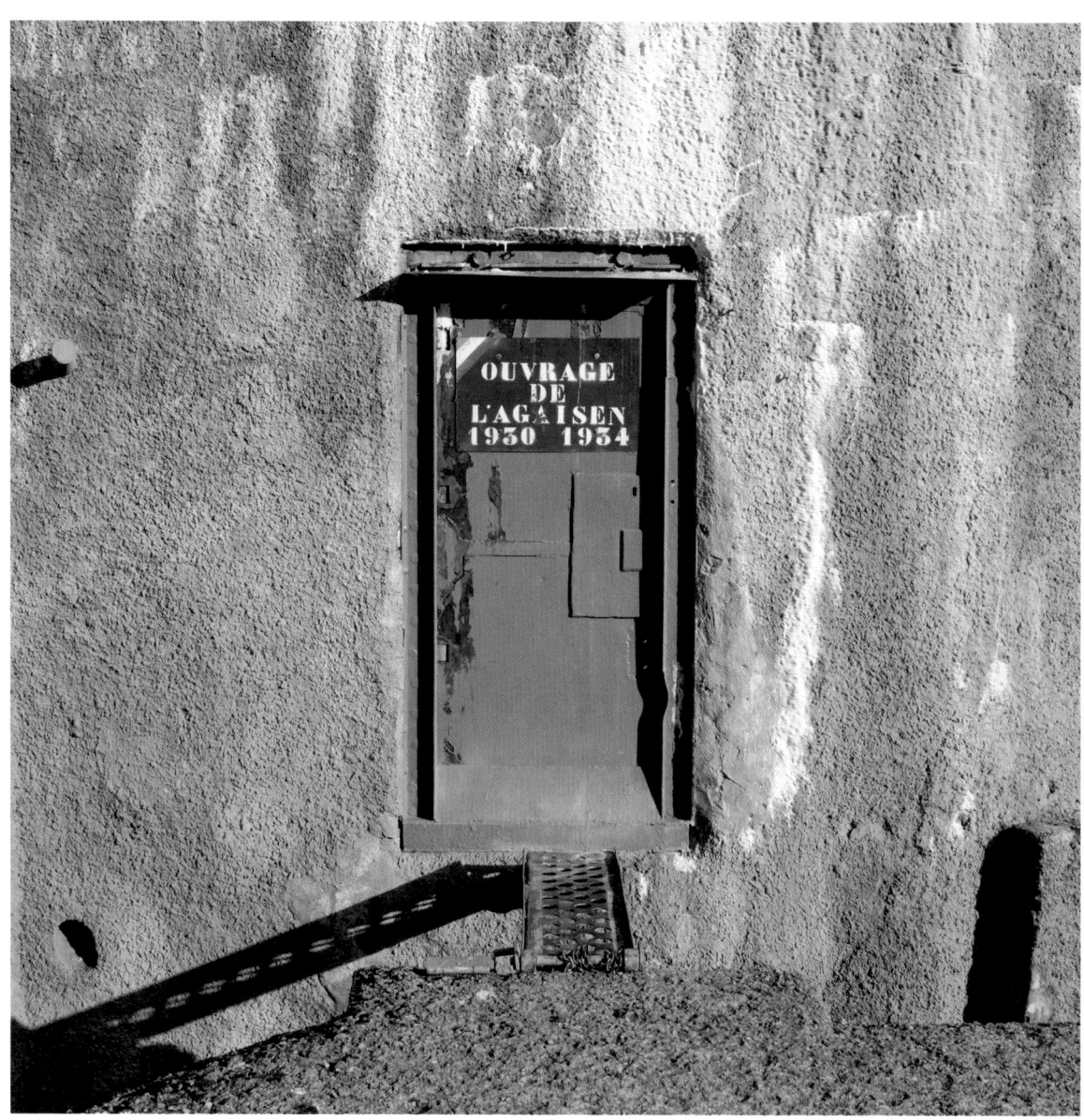

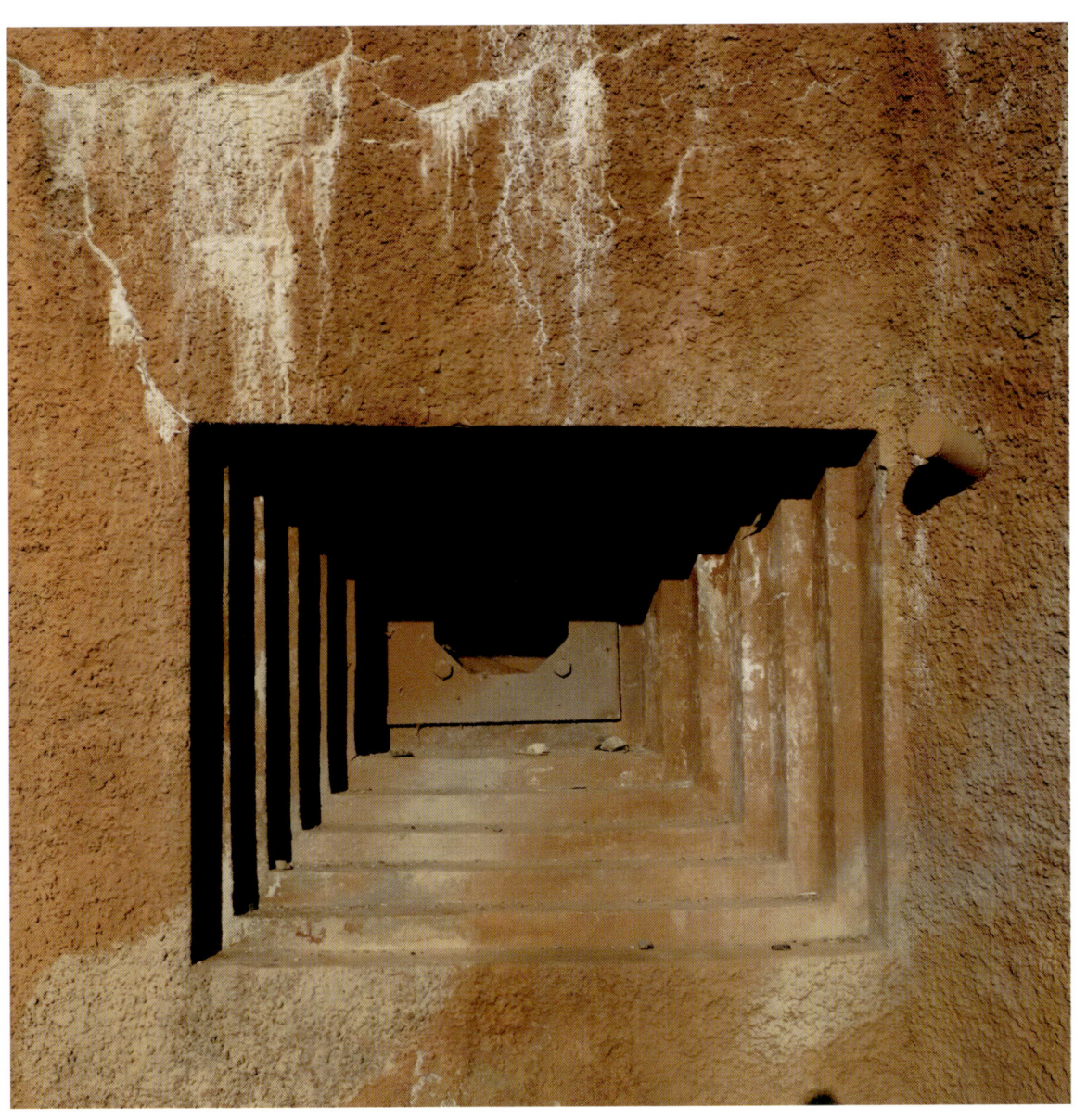

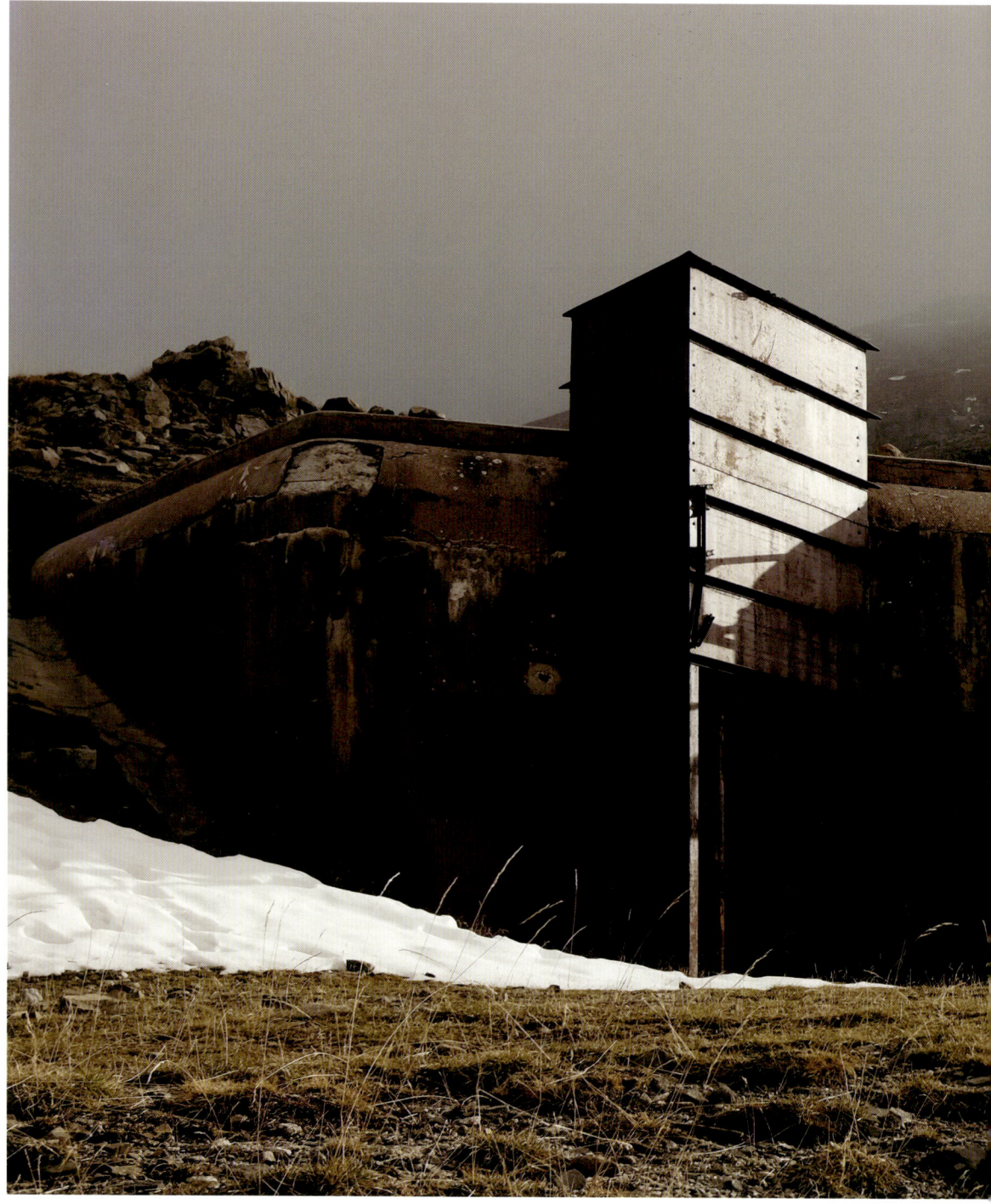

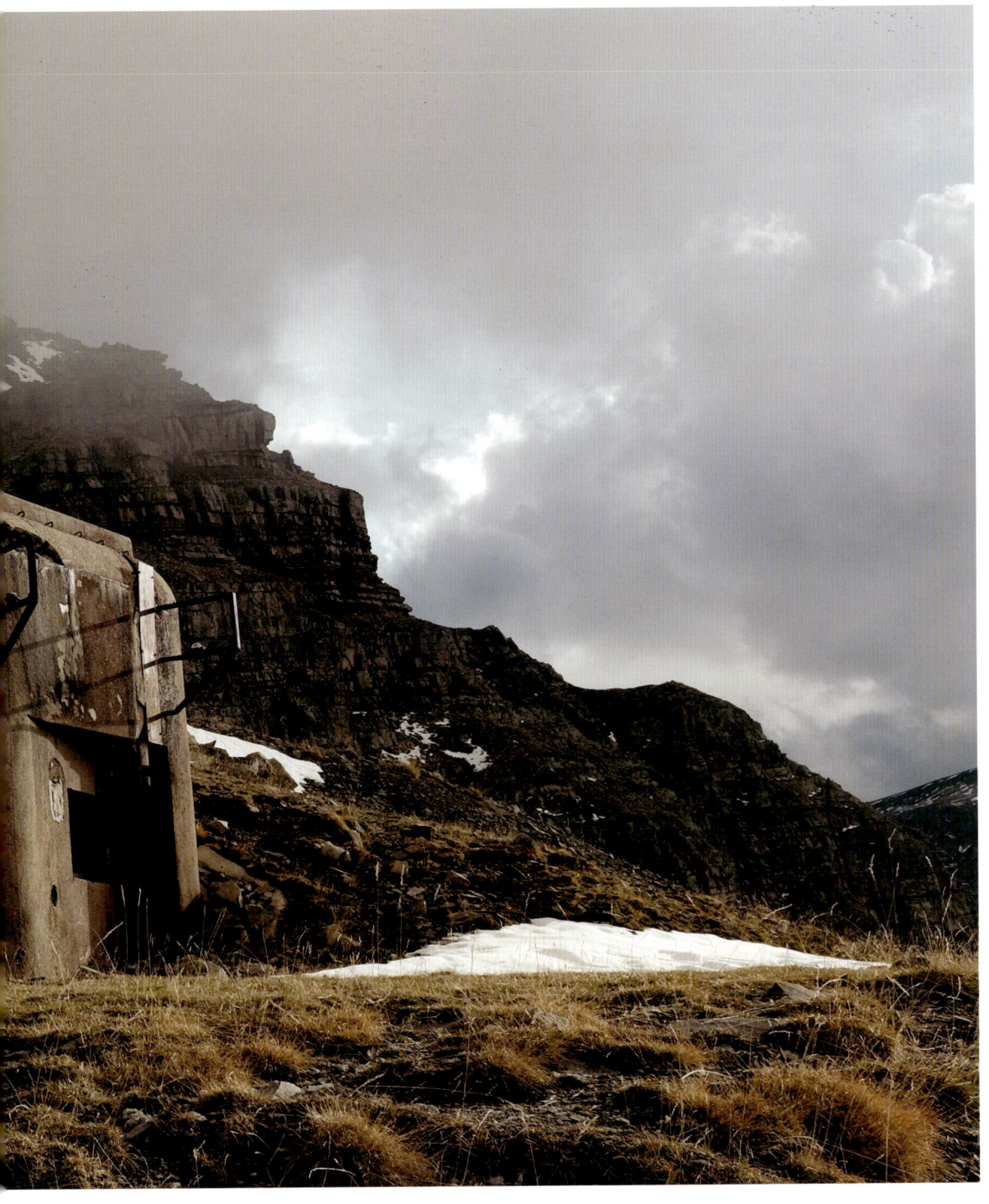

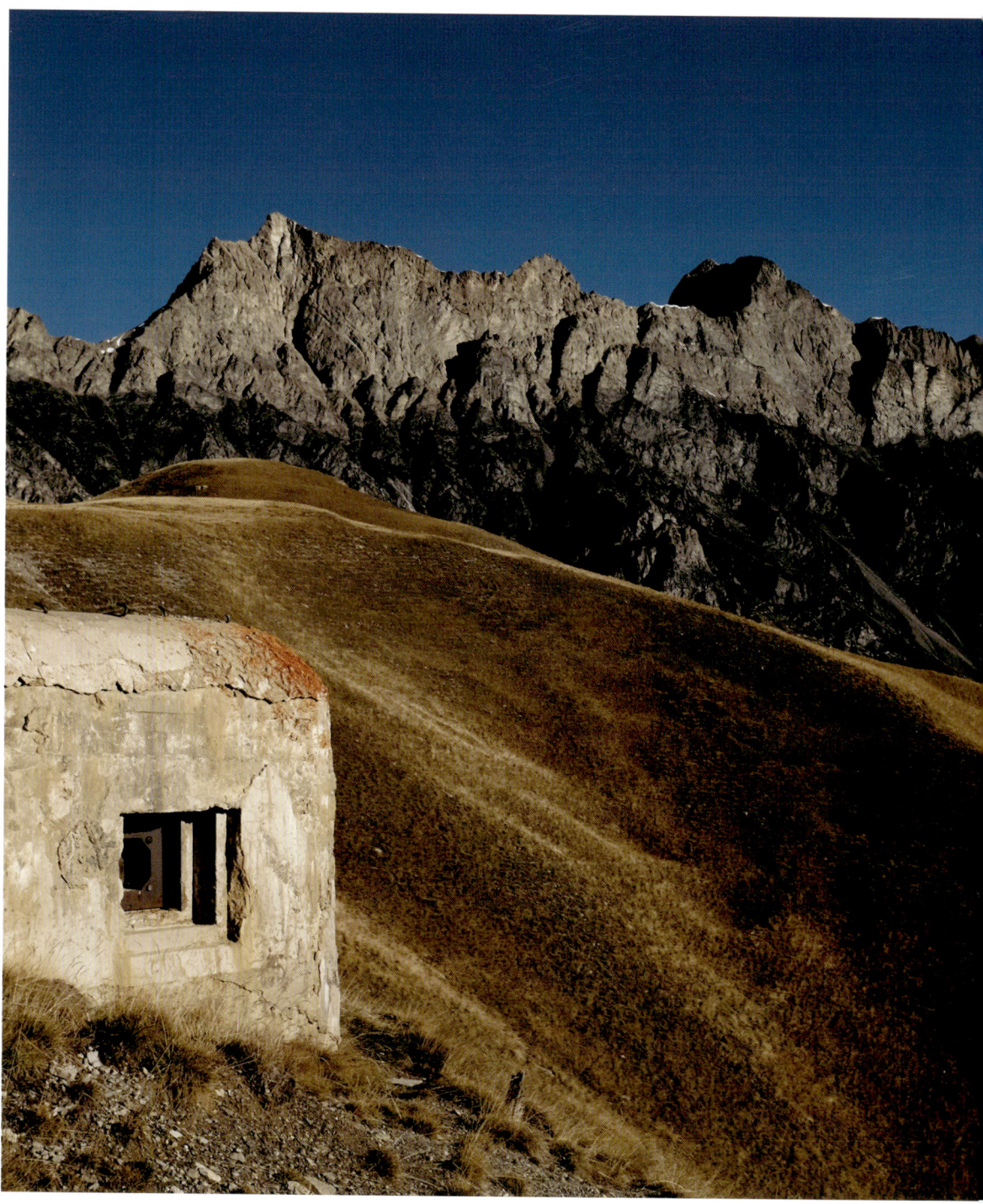

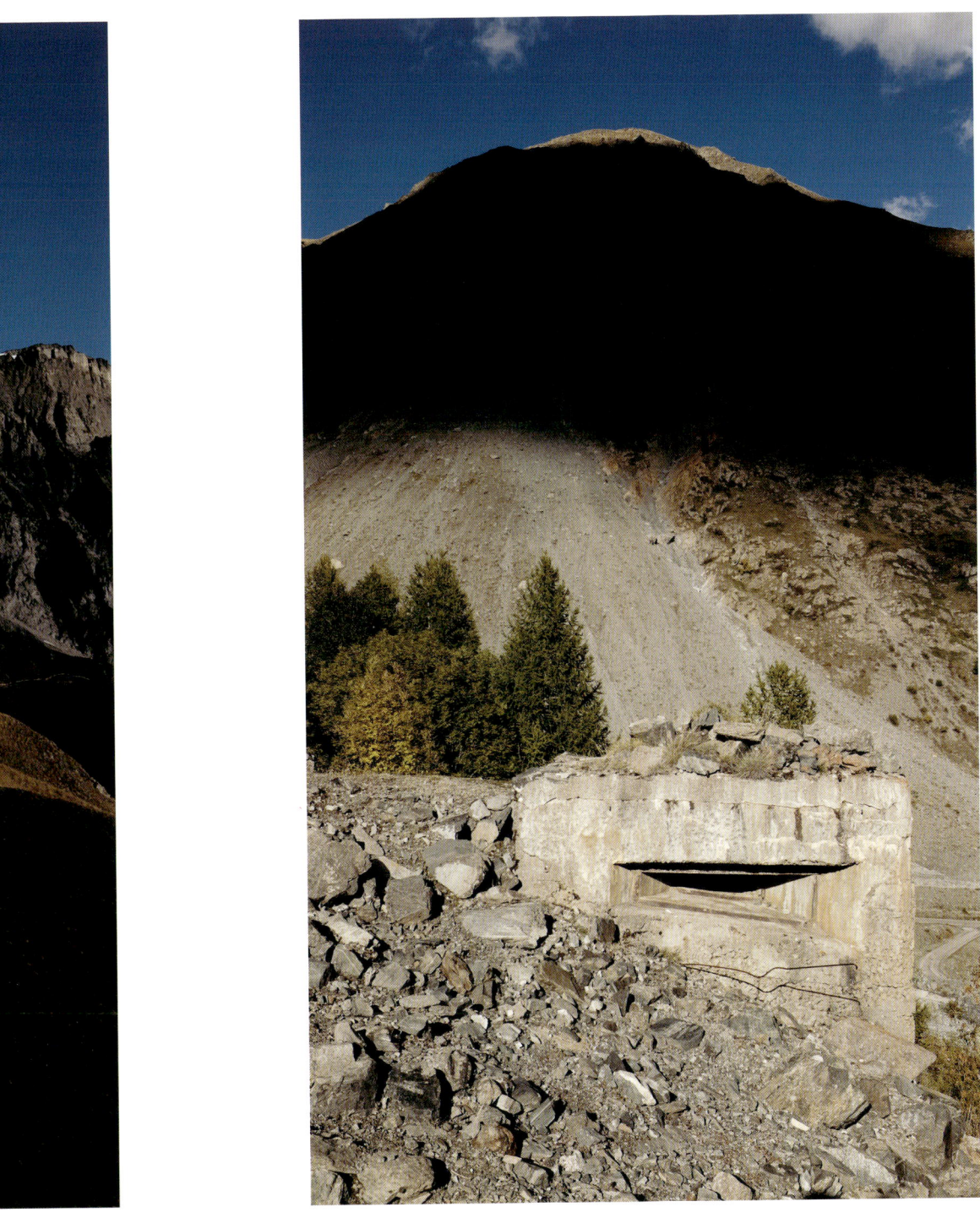

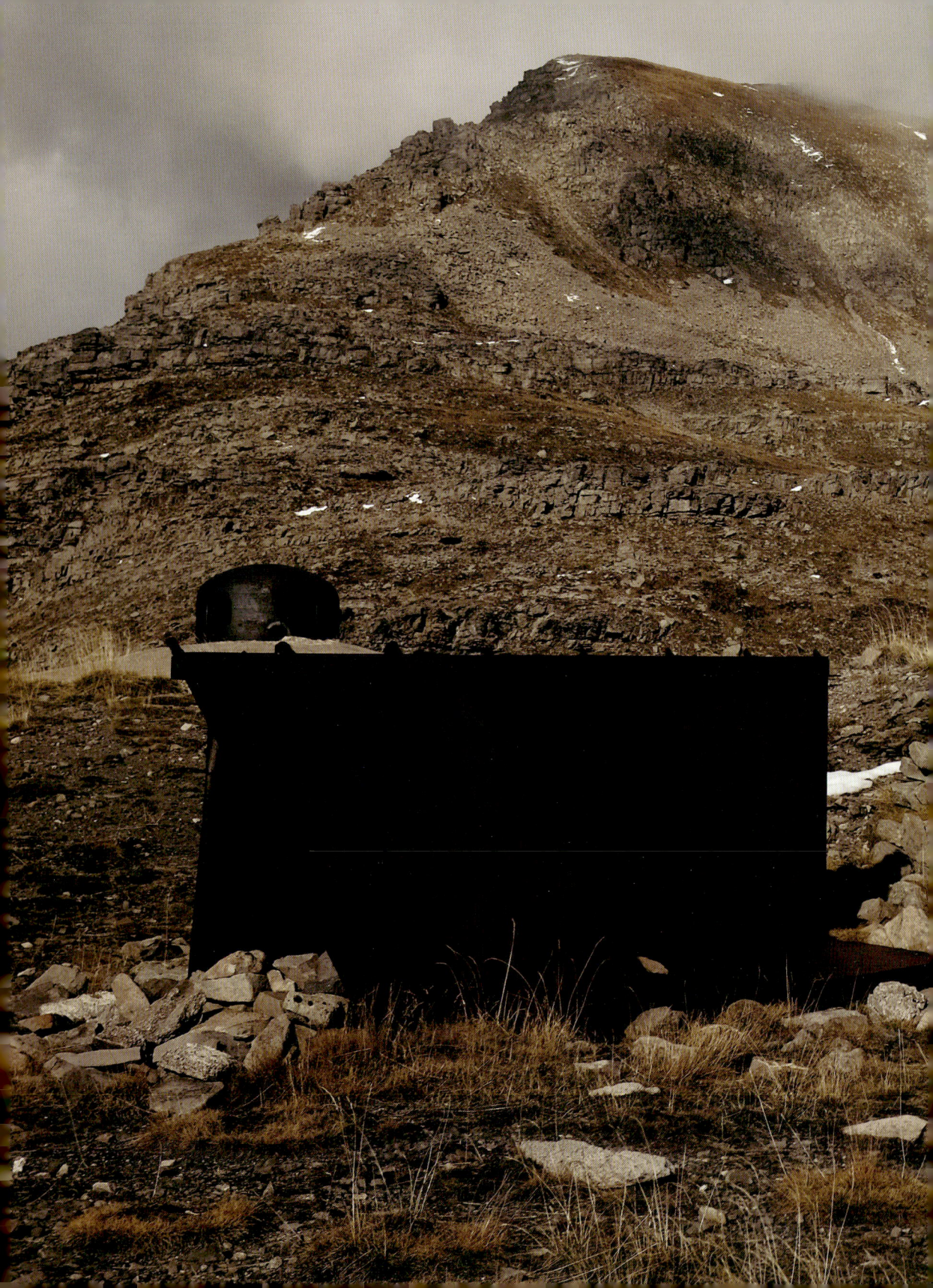

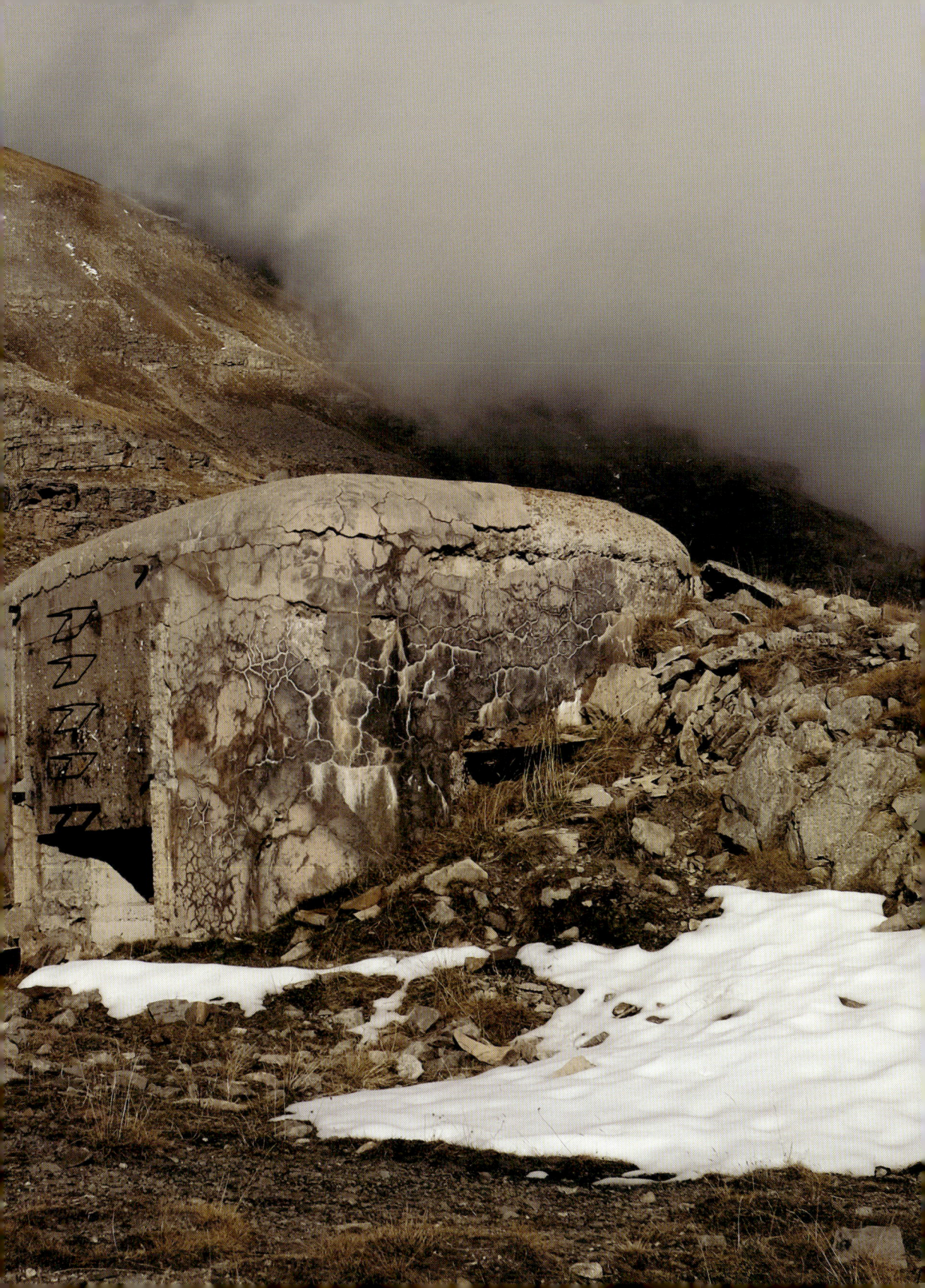

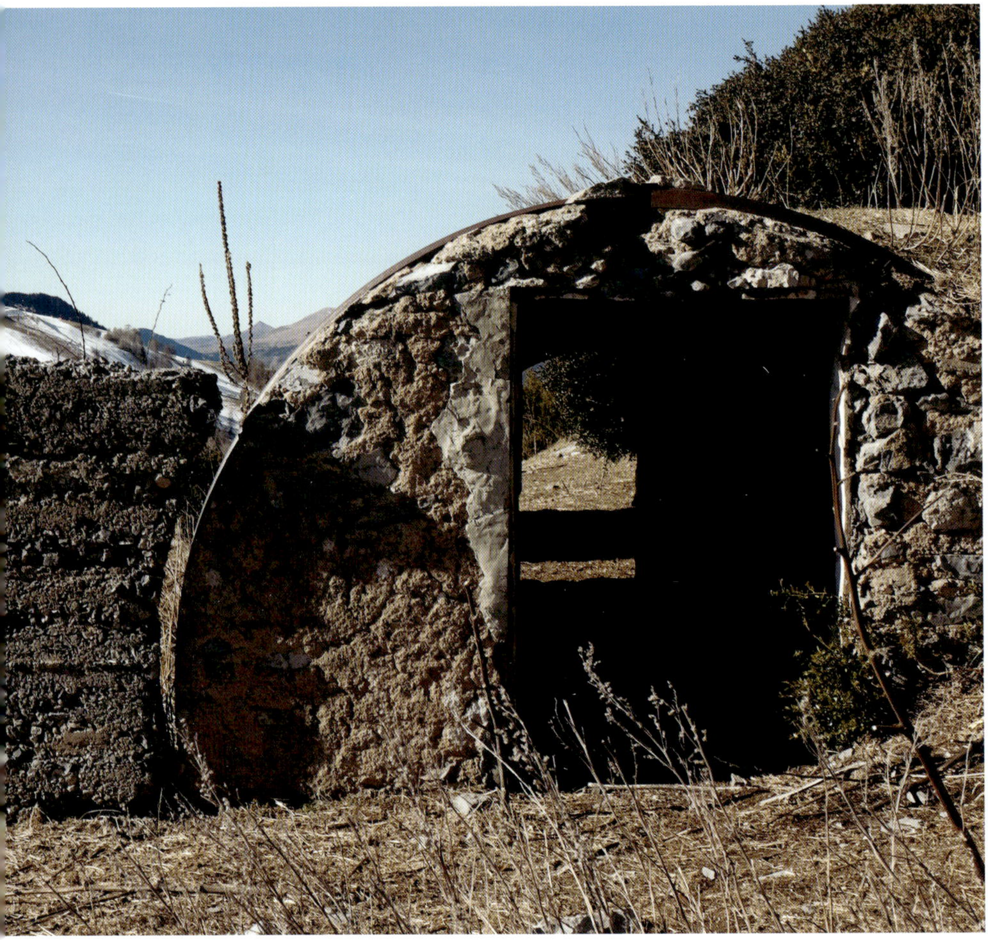

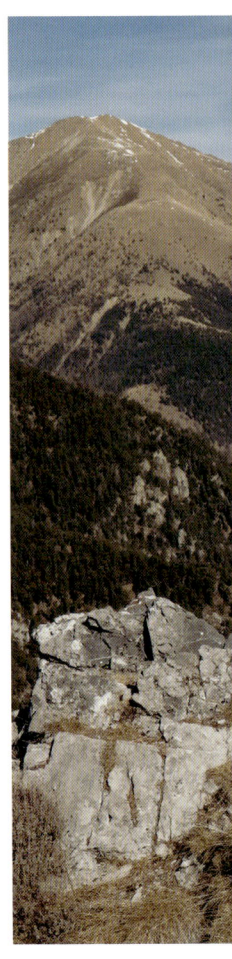

Avant-poste de Conchetas
SFAM / Tinée-Vésubie
One officer, 36 men

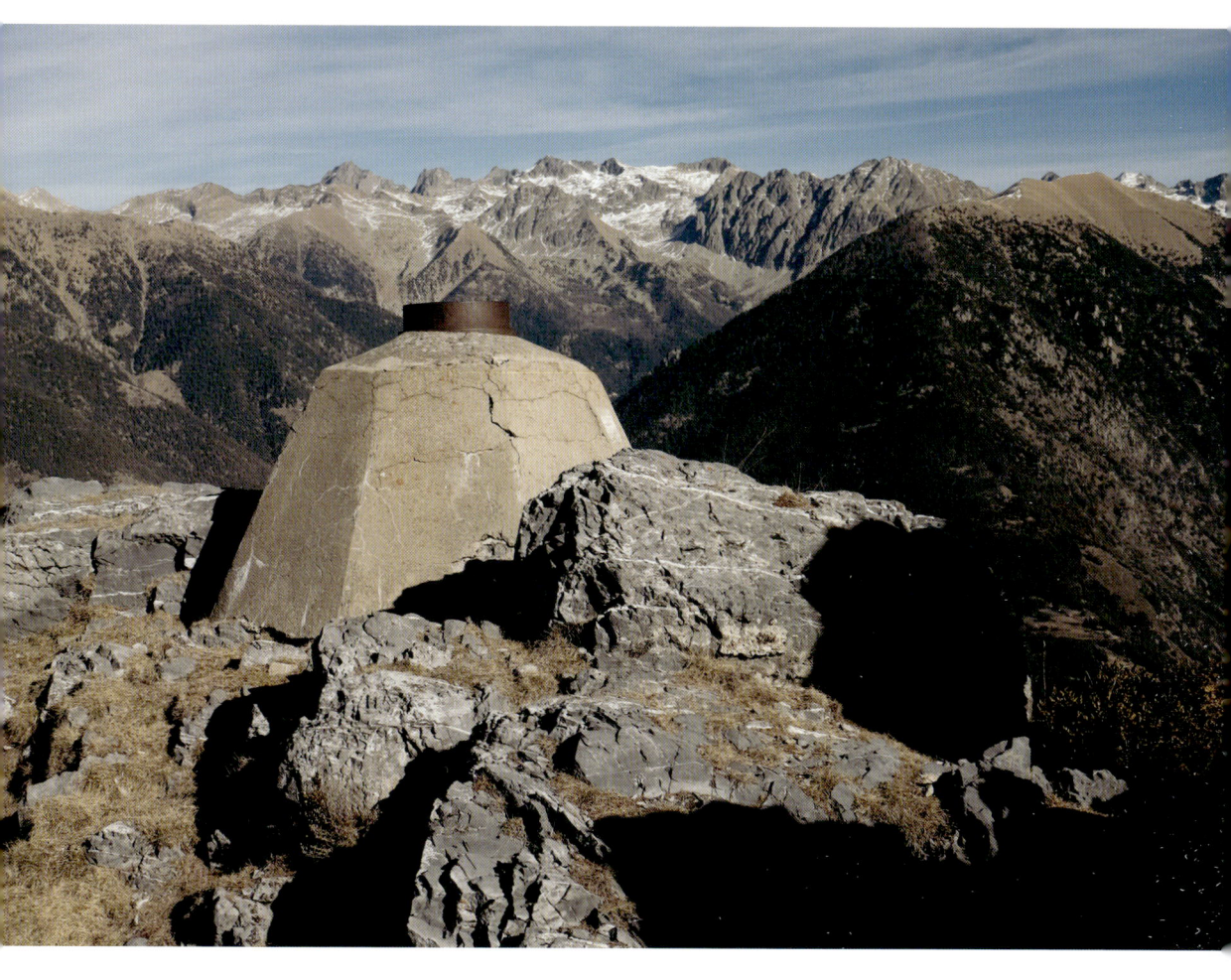

Following pages
Petit ouvrage du Col de Restefond
SFD / Jausiers
One officer, 82 men

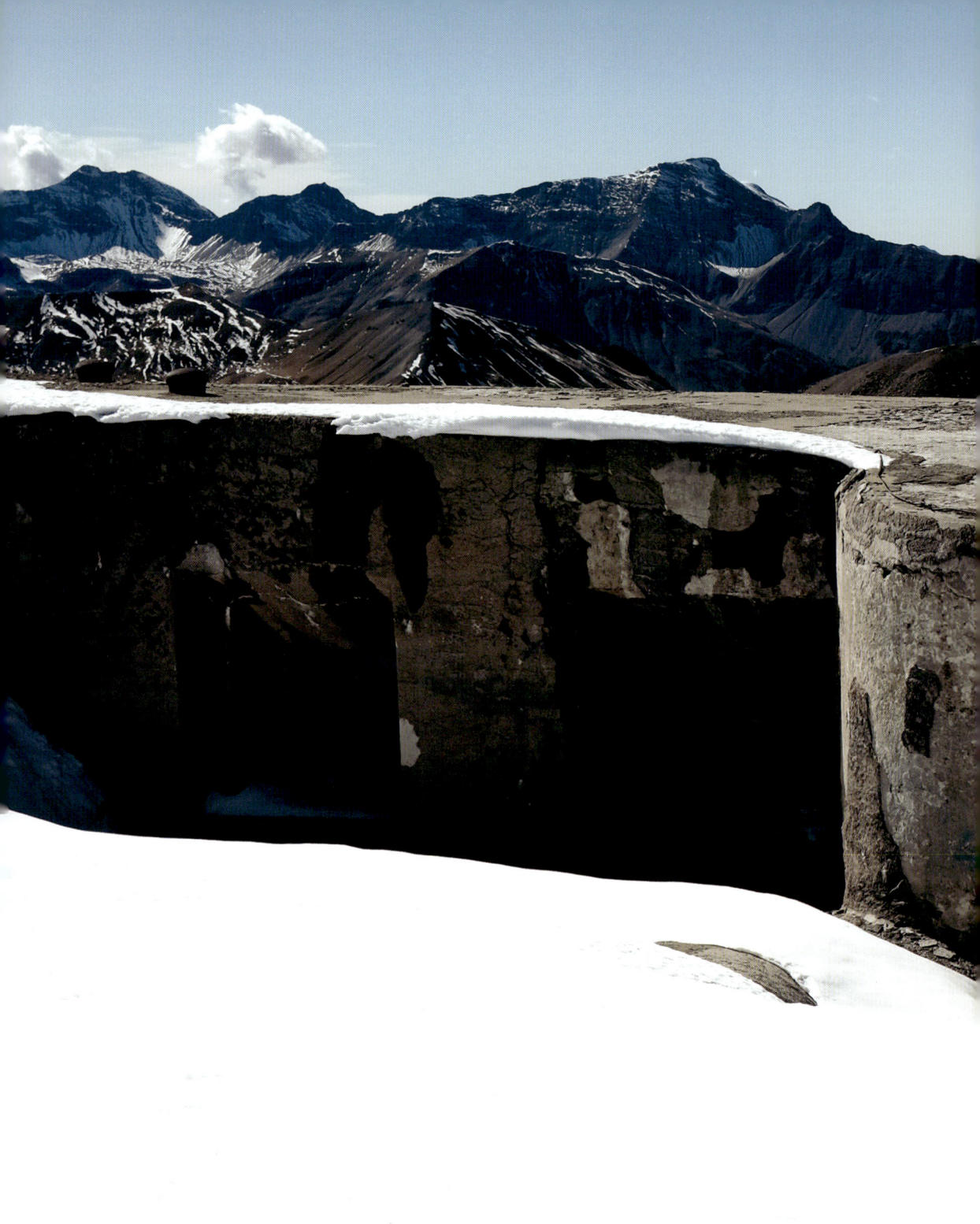

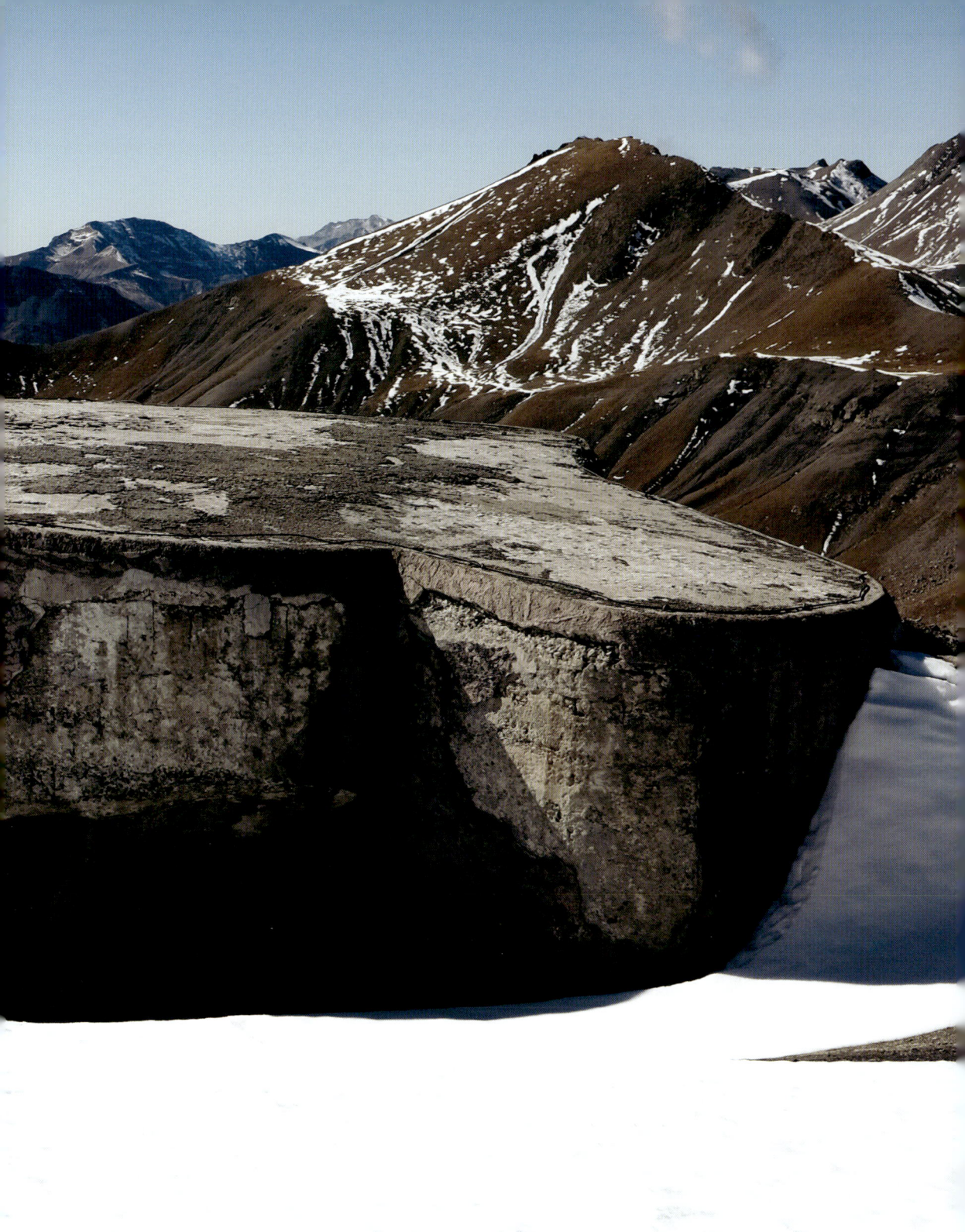

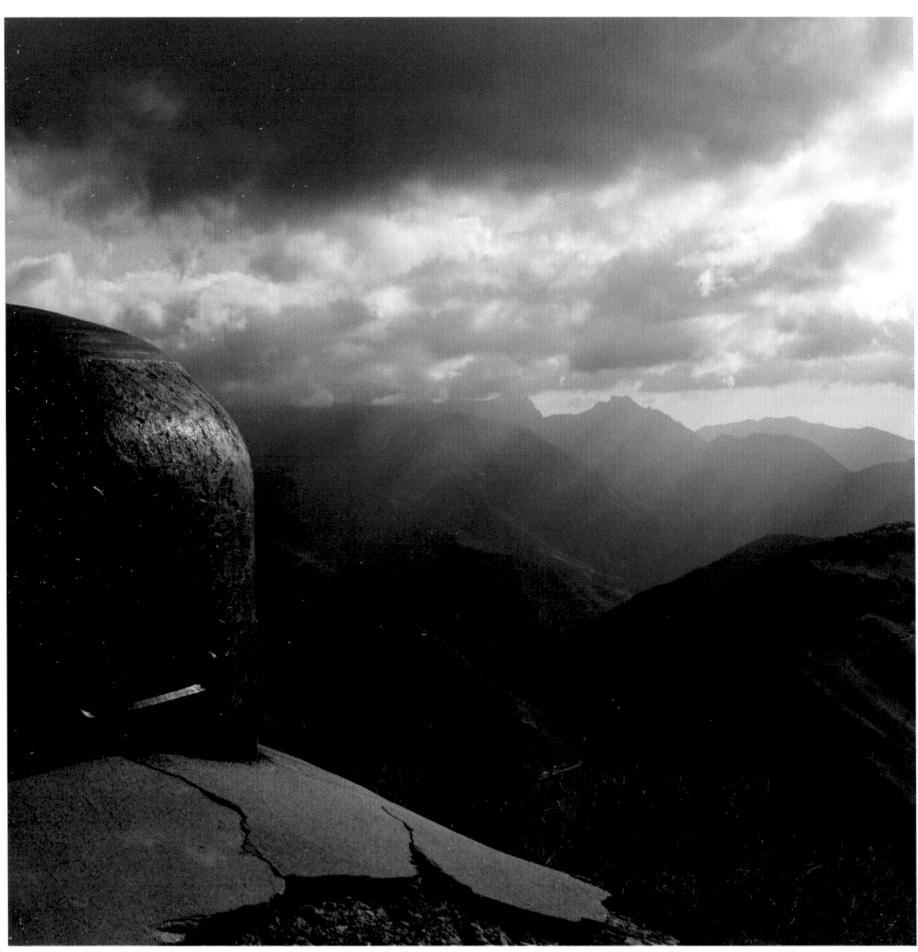

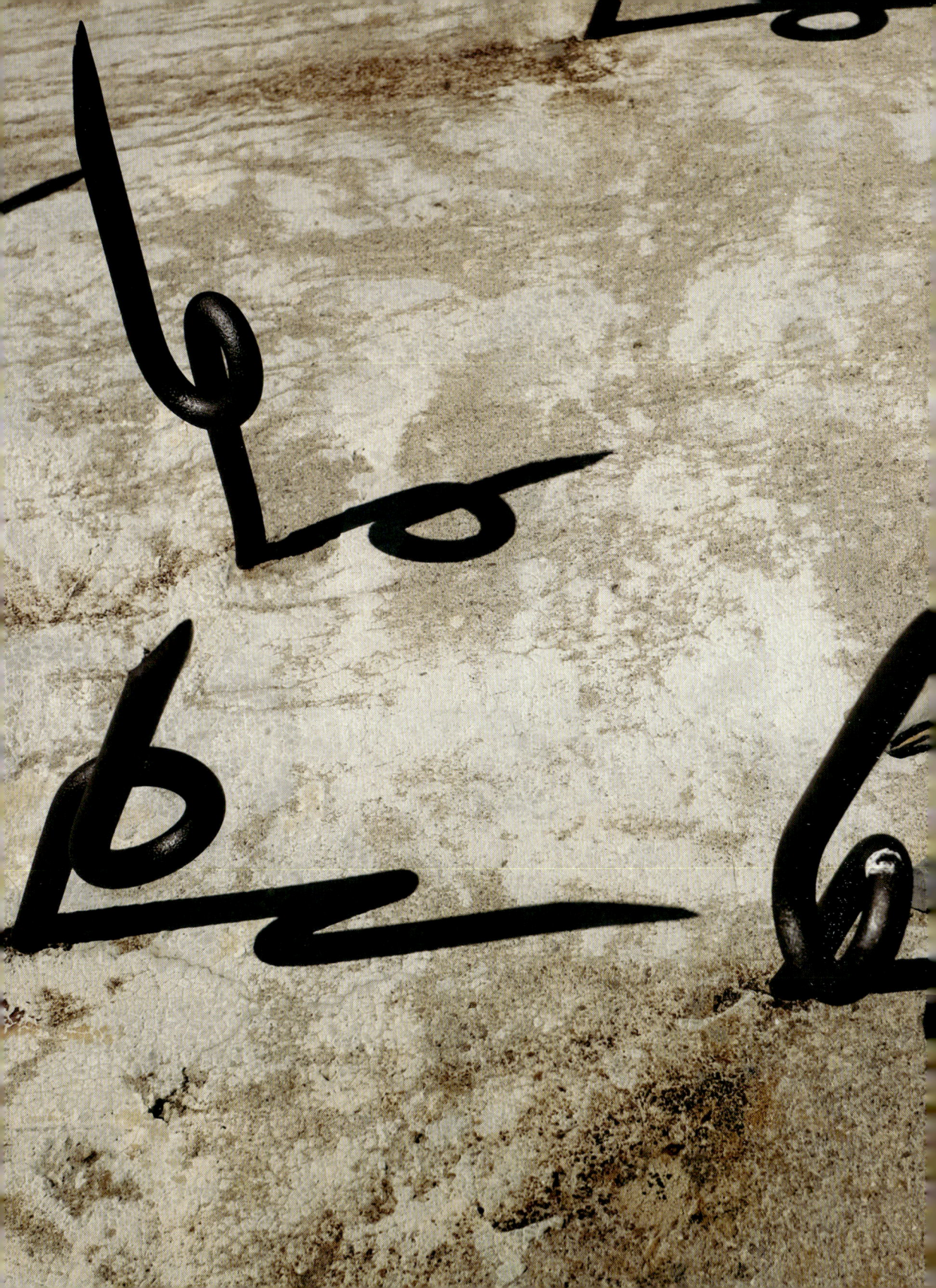

44°11'11"N

7°03'10"E

Bunker Research

Bunkerresearch.com

First published in June 2016 by Isola Press, London
Isolapress.com

Second, expanded edition published in October 2019

Text © Max Leonard
Photography © Camille McMillan
The moral right of the authors has been asserted.

Design: Myfanwy Vernon-Hunt
This-side.co.uk

Printed by EBS in Verona, Italy
ebs-bortolazzi.com

ISBN: 978-0-9954886-4-9
A catalogue record for this book is available from the British Library. All rights reserved. No part of this publication may be reproduced, stored in a retrieval system, or transmitted in any form or by any means, electronic, mechanical, photocopying, recording or otherwise, without the written permission of the publishers.

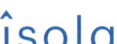